The Biased Brain

A Comprehensive Exploration of Cognitive Distortions

by Bo Bennett, PhD

Formatted, Converted, and Distributed by eBookIt.com
http://www.eBookIt.com

ISBN-13: 9781456641207 (hardcover)
ISBN-13: 9781456641184 (paperback)
ISBN-13: 9781456641177 (ebook)
ISBN-13: 9781456641214 (audiobook)

Dear Esteemed Reader,

Thank you immensely for choosing this book to join your collection. We imagine that you've already embarked on an exploration of ideas within these pages, and we couldn't be happier about it!

Now, if you find yourself chuckling, pondering, or even debating with the words in front of you, we'd absolutely love to hear about it. If you can spare a few moments to pen down your thoughts in a review, we would be as delighted as a dictionary on a spelling bee!

An Amazon review would be excellent - but hey, we're far from picky. Whether it's a scribble on the back of a grocery list, a tweet, or even a message in a bottle (though that might take a while to reach us), your feedback is gold.

Writing a review might not be as fun as a spontaneous dance-off, but we promise it'll bring grins to our faces, warmth to our hearts, and incredibly valuable insights to future readers.

With Gratitude,

Bo Bennett, PhD
Publisher
Archieboy Holdings, LLC.

Table of Contents

Introduction

Welcome! I'm Bo Bennett, PhD, social psychologist and tour guide, navigating the perplexing labyrinth of the human mind. You might know me from my best-selling book, "Logically Fallacious," a merry romp through the world of logical fallacies. But my friends, fallacies are only part of the puzzle, the frosting on the cognitive cake, so to speak. To get a more complete picture of human reasoning (and occasionally, un-reasoning), we need to delve into the realm of cognitive biases. So get ready to have your mind blown–or at least say to yourself, "Wow, that's interesting."

You see, logical fallacies and cognitive biases are like mischievous twins. They look similar and often conspire together, but they're distinct in their own right. Logical fallacies are errors in reasoning, the result of faulty logic. They're akin to reaching an incorrect solution in a math problem because you added when you should have subtracted. Cognitive biases, on the other hand, are systematic errors in our thinking that influence our choices and judgments. They're like a calculator that's always a few digits off, no matter how carefully you punch in the numbers.

"But Bo, how come many of these biases have 'Fallacy' in the name?" Wise observation, my friend. This is because a fallacy, in a general sense of the term, is more synonymous with "error" in the context of language or thought. In that regard, a "logical fallacy" would be a subset of "fallacy". Confused? Don't lose sleep over it. I don't.

Despite their differences, both logical fallacies and cognitive biases are disruptions in our thought process. They prevent us from seeing the world as it is, instead presenting us with a distorted, often self-serving version of reality. It's a bit like wearing rose-tinted glasses that also happen to have smudges, scratches, and the occasional bug splatter.

Take politics, for instance. The Confirmation Bias - a cognitive bias - nudges us to favor information that supports our beliefs, while the Strawman Fallacy - a logical fallacy - misrepresents an opponent's argument to make it easier to attack. Together, they're like a tag team, working to undermine objective reasoning.

In religion, the Belief Bias clouds our judgment of logic and evidence, while the Appeal to Faith Fallacy argues that belief comes from faith, not evidence. These two can tango together, leading to steadfast belief without logical foundation.

And on social issues, such as climate change, the Dunning-Kruger Effect (a cognitive bias) leads people to overestimate their understanding, while the Appeal to Nature Fallacy (a logical fallacy) argues that anything natural is inherently good or right.

In "Cognitively Biased," we'll dive into these cognitive quirks, exploring their impacts, offering countermeasures, and shedding light on the scientific underpinnings. Because to truly understand our minds and to reason effectively, we need to see the whole circus, clowns and all.

So, as we embark on this expedition, remember: understanding is the first step to improvement. With a sprinkle of humor, a dash of insight, and a healthy dose of reality-check, let's make our way through the enthralling world of cognitive biases. Here's to becoming a little less cognitively biased and a whole lot more cognitively savvy!

Actor–Observer Bias

Definition: Actor-Observer Bias is the dandy little tendency we have to blame our own actions on external factors while attributing others' actions to their intrinsic characteristics. It's like the psychological version of wanting to have our cake and eat it too, only in this case, the cake is blame and nobody really wants a slice.

Description: This bias stems from our struggle to see events from a balanced perspective. Think of it as a sort of 'perception see-saw'. On one side, we have the 'actor', where we attribute our own actions to our circumstances rather than our personalities. The 'observer' side, however, sees others' actions as reflective of their character, rather than their situation. These inconsistencies in attribution can lead to some rather unbalanced judgment swings. After all, we all like to think we're the ones dancing with grace while everyone else is just tripping over their own feet.

Examples: To illustrate, consider the realm of politics. Politician A may argue passionately about income inequality, attributing their own wealth to a fortunate upbringing (the actor). Meanwhile, the observer (read: rivals and critics) might attribute Politician A's wealth to inherent greed or corruption.

In religious contexts, imagine a religious leader who falls from grace due to scandal. The leader might attribute their fall to external pressures or temptation (actor), while their followers might see it as a revelation of their true, morally flawed character (observer).

And on social issues, let's consider online trolling. An individual might troll others due to perceived anonymity and lack of consequence (actor), while those on the receiving end might attribute the trolling to the individual's inherent nastiness (observer).

Impact: The Actor-Observer Bias can significantly skew our understanding of others' behaviors and our own. It leads to misunderstanding, feeds conflict, and generally muddies the waters of objective judgment. For society, this bias can perpetuate stereotypes, widen political divisions, and even undermine efforts to promote social justice.

Related Biases: This bias is a cousin to the Fundamental Attribution Error, which is the general tendency to over-emphasize personal characteristics and ignore situational factors when judging others' behavior. It's also a close relative to the Self-Serving Bias, where we credit our successes to our own merit and blame failures on external factors.

Countermeasures: To avoid this bias, we need to play the world's least exciting game of 'switcheroo'. This involves swapping perspectives; trying to see situations from the other person's point of view, and critically examining our own actions as if we were outside observers. Sounds fun, right?

Key Takeaways: The Actor-Observer Bias is like psychological double standards, where we attribute our actions to circumstances and others' actions to their character. It has substantial social impacts, but can be countered by perspective-switching. So the next time you feel the urge to judge, remember to get off the 'perception see-saw' and try a little balance instead.

References:

Malle, B. F. (2006). The actor-observer asymmetry in attribution: A (surprising) meta-analysis. Psychological Bulletin, 132(6), 895–919. https://doi.org/10.1037/0033-2909.132.6.895

Ross, L. (1977). The intuitive psychologist and his shortcomings: Distortions in the attribution process. In L. Berkowitz (Ed.), Advances in experimental social psychology (Vol. 10, pp. 173-220). Academic Press. https://doi.org/10.1016/S0065-2601(08)60357-3

Miller, D. T., & Ross, M. (1975). Self-serving biases in the attribution of causality: Fact or fiction? Psychological Bulletin, 82(2), 213–225. https://doi.org/10.1037/h0076486

Ambiguity Aversion

Definition: Ambiguity Aversion, also known as the Ellsberg Paradox, is our cognitive desire to run screaming from anything uncertain or ambiguous. It's not just a fear of the dark; it's a fear of a slightly shadowy corner in broad daylight.

Description: In essence, ambiguity aversion refers to our preference for known risks over unknown risks. This bias arises from our brain's penchant for certainty and predictability. Uncertainty sends our grey matter into a tizzy, and like a distressed parent losing sight of their child at the supermarket, it starts frantically opting for safer, familiar choices.

Examples: In the world of politics, ambiguity aversion might manifest in our voting behavior. People often vote for the same party or candidate simply because it's a known entity, ignoring potential new candidates who might be better but are less familiar.

Religion is another area where ambiguity aversion is at play. Many individuals stick with the religious beliefs they were raised with, a known risk, instead of exploring other faiths or the lack thereof, an unknown risk.

And let's not forget social issues. People may be averse to new policies or laws that promise improvement but entail a certain degree of uncertainty. For instance, resistance to healthcare reform or new environmental regulations often stems from ambiguity aversion.

Impact: Ambiguity aversion can limit our growth, our understanding, and our progress. It often locks us into the status quo, preventing us from exploring potentially beneficial opportunities, be it in politics, religion, social issues, or daily life. It's like choosing to stay in a mediocre restaurant because we're afraid the one across the street, despite its fantastic reviews, might disappoint us.

Related Biases: Ambiguity aversion is a sibling to the Status Quo Bias, the human tendency to like things to stay relatively the same. It's also kin to Risk Aversion, where we prefer the sure thing over the gamble, even if the gamble might have better potential outcomes.

Countermeasures: Combating ambiguity aversion is like learning to swim in the deep end. It's about gradually increasing our tolerance for uncertainty and practicing decision-making in ambiguous situations. It can help to gather as much information as possible, weigh the pros and cons, and then take the plunge.

Key Takeaways: Ambiguity aversion is our cognitive distaste for uncertainty, which can limit our openness to new experiences and potentially beneficial opportunities. It's a tricky bias to overcome, but with patience, information, and a willingness to cannonball into the deep end of ambiguity, we can begin to navigate uncertainty with a little more finesse.

References:

Ellsberg, D. (1961). Risk, Ambiguity, and the Savage Axioms. The Quarterly Journal of Economics, 75(4), 643-669. https://doi.org/10.2307/1884324

Samuelson, W., & Zeckhauser, R. (1988). Status Quo Bias in Decision Making. Journal of Risk and Uncertainty, 1, 7-59. https://doi.org/10.1007/BF00055564

Kahneman, D., & Tversky, A. (1979). Prospect Theory: An Analysis of Decision under Risk. Econometrica, 47(2), 263-291. https://doi.org/10.2307/1914185

Ambiguity Effect

Definition: The Ambiguity Effect is a cognitive bias where we shun options with missing information, even if they're potentially more rewarding. It's the feeling you get when you opt for an apple over an exotic fruit because, well, who knows what that thing tastes like?

Description: The Ambiguity Effect, a cousin to the Ambiguity Aversion, emerges from our desire for certainty and predictability. We tend to go for options where we have more information, even if it means potentially missing out on something better. It's like picking a predictable movie over a potentially groundbreaking one, just because you know the actor in the first one.

Examples: In politics, the Ambiguity Effect can lead to people shying away from voting for a promising candidate with less public exposure. They might opt for a well-known candidate with a less stellar record, simply because there's more information available about them.

In religion, the Ambiguity Effect might stop someone from exploring a different faith or philosophy. The uncertainty about the new belief system can make their known faith seem more appealing, even if it might not be satisfying them spiritually.

On social issues, let's say there's a new but vaguely described policy on education reform. Despite its potential to dramatically improve education standards, people might oppose it simply because the plan lacks detailed information.

Impact: The Ambiguity Effect can make us miss out on potentially beneficial opportunities, stifle innovation, and promote a fear-based decision-making process. It encourages playing it safe, even when playing it safe means staying stuck in less-than-optimal situations.

Related Biases: The Ambiguity Effect shares a common thread with the Status Quo Bias and Risk Aversion. It's also related to the Uncertainty Effect, where our dislike for uncertainty affects our decision-making process.

Countermeasures: To fight the Ambiguity Effect, we need to learn to be comfortable with a certain degree of uncertainty. This involves stepping outside of our comfort zones, seeking out new information, and evaluating options based on potential benefits, not just available information.

Key Takeaways: The Ambiguity Effect is our tendency to shy away from uncertainty, even at the cost of potential benefits. By embracing a bit of uncertainty and making decisions based on potential outcomes, we can navigate life a bit more adventurously and potentially more fruitfully.

References:

Ellsberg, D. (1961). Risk, Ambiguity, and the Savage Axioms. The Quarterly Journal of Economics, 75(4), 643-669. https://doi.org/10.2307/1884324

Kahneman, D., & Tversky, A. (1979). Prospect Theory: An Analysis of Decision under Risk. Econometrica, 47(2), 263-291. https://doi.org/10.2307/1914185

Fox, C. R., & Tversky, A. (1995). Ambiguity Aversion and Comparative Ignorance. The Quarterly Journal of Economics, 110(3), 585-603. https://doi.org/10.2307/2946693

Anchoring Bias

Definition: Anchoring Bias is our cognitive tendency to rely heavily on the first piece of information (the "anchor") we receive when making decisions. Picture it as a mind-anchor stubbornly sticking to the seabed while you're trying to navigate your boat - the boat being your decision-making process.

Description: The Anchoring Bias is a peculiar quirk of the human mind. We tend to give undue weight to the initial information we receive and then adjust our subsequent thoughts and judgments based on this anchor. It's like trying to estimate the value of a used car based on its original price tag, ignoring factors like its current condition and mileage.

Examples: In politics, initial poll results often act as anchors, affecting how people perceive the candidates. Once we form an opinion based on these polls, we might ignore later information that contradicts our initial impressions.

In religion, the first religious beliefs or practices we're exposed to often become anchors. They heavily influence our perception of other faiths and philosophies.

When discussing social issues, the first piece of news or opinion we come across can act as an anchor, coloring our perception and interpretation of subsequent information about the issue.

Impact: The Anchoring Bias can lead us to make less-than-optimal decisions and judgments. It can skew our perception of people, ideas, and events, potentially blinding us to more accurate or helpful information that comes later.

Related Biases: Anchoring Bias is a close cousin of the Confirmation Bias, where we seek out information that confirms our existing beliefs. It also shares a family resemblance with the Availability Heuristic, which leads us to overestimate the importance of information that's readily available or memorable.

Countermeasures: To counteract the Anchoring Bias, consciously seek diverse information before making a decision. Don't let the first piece of information dictate your perspective. And remember, your mind-anchor is not made of adamantium; it can and should be lifted when needed.

Key Takeaways: Anchoring Bias is our cognitive tendency to stick too closely to our first impressions. By recognizing this bias and actively seeking out diverse and later information, we can make more balanced and informed decisions.

References:

Tversky, A., & Kahneman, D. (1974). Judgment under Uncertainty: Heuristics and Biases. Science, 185(4157), 1124–1131. https://doi.org/10.1126/science.185.4157.1124

Kahneman, D. (2011). Thinking, Fast and Slow. Farrar, Straus and Giroux.

Furnham, A., & Boo, H. C. (2011). A literature review of the anchoring effect. The Journal of Socio-Economics, 40(1), 35-42. https://doi.org/10.1016/j.socec.2010.10.008

Anthropic Bias

Definition: Anthropic Bias is our tendency to reason as if our particular moment in history, our particular location, or our particular species is somehow central or typical in the grand scheme of things. It's like thinking you're the center of the universe, and let's be real, who hasn't entertained that thought at some point?

Description: Anthropic Bias can distort our perception of reality by making us presume our perspective is "the norm". It arises from our difficulty to detach ourselves from our personal experiences and limited perspectives when trying to understand the universe and its laws. It's like watching a soccer game only through the GoPro strapped to your favorite player's head.

Examples: In politics, a country might assume that its form of government is the most effective one, ignoring the successes and nuances of others. This can lead to flawed foreign policies and diplomatic misunderstandings.

In religion, anthropic bias can lead to the belief that humans are the purpose or focal point of the universe or divine creation. This can influence how we perceive and interact with nature and other species.

In social issues, anthropic bias can lead to an inability to understand or respect cultures, values, or practices different from ours. For instance, a person might find it hard to comprehend why certain societies uphold certain traditions, simply because it's far removed from their own experiences.

Impact: Anthropic Bias can hamper our understanding of the universe, constrain our thinking, and fuel misunderstanding and conflict. It can hinder our ability to grasp the diversity and complexity of the world around us.

Related Biases: The anthropic bias is closely related to the False Consensus Effect, where we assume our opinions are representative of the majority. Egocentric Bias, where we overly rely on our own perspective, is another close relative.

Countermeasures: Mitigating the Anthropic Bias requires a conscious effort to broaden our perspectives. We need to actively seek out diverse viewpoints, experiences, and cultures. Recognize that our viewpoint is one among many, and that the universe doesn't revolve around our tiny existence.

Key Takeaways: Anthropic Bias reflects our tendency to see the world from our limited and subjective vantage point. By embracing a more humble, curiosity-driven approach, we can counteract this bias and enrich our understanding of the universe and its diversity.

References:

Bostrom, N. (2002). Anthropic Bias: Observation Selection Effects in Science and Philosophy. Routledge.

Page, S. E. (2007). The Difference: How the Power of Diversity Creates Better Groups, Firms, Schools, and Societies. Princeton University Press.

Appeal to Novelty

Definition: The Appeal to Novelty is our cognitive predisposition to favor newer ideas, innovations, or things simply because they're new. It's like a child with a shiny new toy, except we're all adults and the toys are concepts and objects.

Description: The Appeal to Novelty is rooted in our attraction to novelty, excitement, and discovery. We like to think that the new must be improved. And while that's often the case, it's not always true. Like assuming your new smartphone must be better, until you realize the battery life is worse than an antique pocket watch.

Examples: In politics, we often see the Appeal to Novelty at play when voters get excited about a fresh-faced candidate promising groundbreaking reforms, despite a lack of substantive policies or experience.

In religion, the emergence of new spiritual movements or reinterpretations of ancient teachings can attract followers, drawn by the allure of the "new."

In social issues, novel solutions to long-standing problems may be touted as the best without sufficient evidence. For instance, adopting new educational technologies without proper evaluation, just because they're "the future."

Impact: The Appeal to Novelty can lead us to make choices that aren't necessarily the best. We might adopt newer, untested technologies, or discard established, effective strategies merely because they're "old."

Related Biases: This bias is an antithesis to the Appeal to Tradition, where old is seen as gold. It also shares its lineage with the Bandwagon Effect, where popularity dictates our preferences.

Countermeasures: To combat the Appeal to Novelty, we should evaluate the merits of ideas, technologies, or strategies based on evidence, not their age. Always remember, new doesn't automatically mean better.

Key Takeaways: The Appeal to Novelty reflects our fondness for the "new." By valuing evidence and substance over novelty, we can make more informed, rational decisions.

References:

Baron, J. (2008). Thinking and Deciding, 4th ed. Cambridge University Press.

Kahneman, D. (2011). Thinking, Fast and Slow. Farrar, Straus and Giroux.

Meyer, R. (2014). The Appeal of the New: A bias in our decision making? In B. Todd (Ed.), How to decide what to do (pp. 77–84). CreateSpace Independent Publishing Platform.

Argument from Repetition (Argument Ad Nauseam)

Definition: The Argument from Repetition is a cognitive bias where repeated statements are perceived as more truthful than they actually are. In other words, if you say something enough times, it starts to sound like a catchy song you can't get out of your head, or worse, it begins to sound true.

Description: The power of repetition can't be underestimated. The underlying mechanism here is the 'illusory truth effect,' which describes our tendency to believe information to be correct after repeated exposure. This bias acts like an annoying dinner guest who won't stop talking about their pet theory. You might not agree at first, but after the fifth explanation, you begin to question your sanity and their point starts to make an eerie amount of sense.

Examples: In politics, repetitive campaign slogans ("Make America Great Again") can create a perceived truth among constituents, regardless of the factual basis.

Religion often uses repetition, with prayers, mantras, or doctrines repeated until they are accepted without question.

Socially, stereotypes can perpetuate through society via repetitive assertions, regardless of their truth.

Impact: This bias can distort our perception of truth, swaying public opinion, reinforcing stereotypes, and creating dogmatic beliefs. You might end up voting for a candidate you otherwise wouldn't or believing in a stereotype that doesn't hold water.

Related Biases: The Continued Influence Effect, where discredited information continues to influence our thinking, and the Mere Exposure Effect, where we develop a preference for things merely because they're familiar, are close cousins.

Countermeasures: Healthy skepticism and fact-checking can keep this bias in check. Just because something is repeated doesn't make it true.

Key Takeaways: The Argument from Repetition demonstrates the power of repetition in shaping our beliefs. Remember, truth is not a function of repetition.

References:

Hasher, L., Goldstein, D., & Toppino, T. (1977). Frequency and the conference of referential validity. Journal of Verbal Learning and Verbal Behavior, 16(1), 107-112.

Fazio, L. K., Brashier, N. M., Payne, B. K., & Marsh, E. J. (2015). Knowledge does not protect against illusory truth. Journal of Experimental Psychology: General, 144(5), 993–1002.

Schwarz, N., Sanna, L. J., Skurnik, I., & Yoon, C. (2007). Metacognitive experiences and the intricacies of setting people straight: Implications for debiasing and public information campaigns. Advances in Experimental Social Psychology, 39, 127–161.

Association Fallacy

Definition: The Association Fallacy, or guilt by association, is a cognitive bias where the qualities of one thing are inaccurately attributed to another due to their association. It's like assuming your neighbor's cat must be evil because it hangs out with the raccoons.

Description: This bias emerges from our penchant to create mental shortcuts, or heuristics. We tend to believe that things, people, or ideas associated with something negative (or positive) must also share those qualities, kind of like the dubious assumption that wearing a lab coat automatically makes one a genius.

Examples:

In politics, politicians are often either praised or criticized based on the people they associate with, rather than their actual policies.

In religion, one may be viewed as a "sinner" simply by associating with individuals deemed to be such by a religious community.

In social issues, if a person hangs out with a group known for a negative stereotype, they may be unjustly attributed those same characteristics.

Impact: This bias can lead to unfair judgments, stereotypes, and discrimination. It can create guilt by association and can promote the spread of misinformation.

Related Biases: The Halo Effect, where one positive attribute is generalized to all attributes of an individual, and the Horn Effect, its negative counterpart, are siblings of the Association Fallacy.

Countermeasures: Critical thinking and not jumping to conclusions can mitigate this bias. Remember, correlation does not mean causation.

Key Takeaways: The Association Fallacy reminds us that associations can be misleading, and qualities don't always transfer by association. Keep your judgments based on individual merits, not company kept.

References:

McElroy, T., & Seta, J. (2003). Framing effects: An analytic–holistic perspective. Journal of Experimental Social Psychology, 39(6), 610-617.

Tversky, A., & Kahneman, D. (1974). Judgment under Uncertainty: Heuristics and Biases. Science, 185(4157), 1124–1131.

Association-Transfer Bias

Definition: The Association-Transfer Bias is a cognitive bias where emotions linked to one stimulus are transferred to another through association. It's a bit like getting mad at your office chair because your boss yelled at you while you were sitting on it.

Description: This bias stems from the way our brains process information. We create connections between stimuli and the emotions they evoke, sometimes resulting in the emotions linked to one stimulus being applied to another, unrelated one. It's like the blame game in our brains, where unrelated things are blamed for emotions they didn't cause.

Examples: In the realm of religion, an individual might associate negative feelings towards a particular belief system because of a negative experience with a single adherent.

In politics, a scandal involving a politician from one party might result in negative emotions being transferred to all politicians from that party.

On social issues, if a person has a negative experience with an individual from a certain demographic, they might transfer these negative emotions to the entire group, fueling stereotypes and prejudices.

Impact: The Association-Transfer Bias can contribute to misunderstandings, stereotyping, and discrimination. It can also impact relationships and decision-making processes negatively.

Related Biases: The Association Fallacy and the Halo Effect bear a family resemblance to the Association-Transfer Bias.

Countermeasures: Being aware of the bias and engaging in mindful thinking can help keep this bias in check. Always analyze the root of your emotions to ensure they're directed at the right source.

Key Takeaways: Association-Transfer Bias is a good reminder that misplaced emotions can distort our perspective. So, the next time you're upset with your boss, remember it's not your chair's fault.

References:

Gawronski, B., & Bodenhausen, G. V. (2006). Associative and propositional processes in evaluation: An integrative review of implicit and explicit attitude change. Psychological Bulletin, 132(5), 692–731.

Zajonc, R. B. (1980). Feeling and thinking: Preferences need no inferences. American Psychologist, 35(2), 151–175.

Attentional Bias

Definition: Attentional Bias is the brain's tendency to pay more attention to certain stimuli, somewhat like a picky eater who only sees the broccoli on a plate full of delicacies.

Description: This cognitive bias involves our attention being disproportionately drawn to certain factors while others are ignored or marginalized. It's like our mind has a spotlight, and whatever it illuminates gets all the focus, regardless of whether it deserves it or not.

Examples: In religious contexts, an adherent may focus on passages that support their belief and ignore ones that present challenges, leading to a skewed understanding of the text.

Politically, people might pay more attention to news items that align with their views and disregard information that contradicts them. It's like reading only every other line of a story and claiming to understand it fully.

In social issues, someone might concentrate on negative news about a particular demographic, fostering an unfair bias against the entire group.

Impact: Attentional Bias can lead to a narrow view of the world, as well as polarized opinions and ignorance of broader contexts. It's a dream come true for misinformation and a nightmare for nuanced understanding.

Related Biases: Confirmation Bias and Selection Bias share a bunk bed with Attentional Bias in the dormitory of cognitive biases.

Countermeasures: Cultivate awareness of the bias and intentionally seek out diverse information sources. It might feel like forcing yourself to eat broccoli, but it's good for you.

Key Takeaways: Attentional Bias reminds us that what we focus on shapes our perception and understanding. It's worth questioning if our spotlight is casting light on the whole picture or just a fragment.

References:

MacLeod, C., Mathews, A., & Tata, P. (1986). Attentional bias in emotional disorders. Journal of Abnormal Psychology, 95(1), 15–20.

Eysenck, M. W., Derakshan, N., Santos, R., & Calvo, M. G. (2007). Anxiety and cognitive performance: Attentional control theory. Emotion, 7(2), 336–353.

Attentional Blink

Definition: The Attentional Blink is like the brain's version of an eye blink - a brief mental lapse in which we miss important details because we're busy processing something else.

Description: This cognitive hiccup typically happens when two pieces of information are presented in quick succession. Our brain gets so caught up with the first piece that it dozes off for a split second, often missing the second piece.

Examples: In religious debates, a flurry of arguments may be thrown around. If one point is particularly intriguing or challenging, the next point may go completely unnoticed, like a fly by night.

In the political sphere, during a fast-paced debate, a candidate might make a great point, but if they immediately follow it with another, the audience might be too busy digesting the first point to catch the second. It's like trying to catch two balls thrown at you simultaneously.

For social issues, a person might miss critical aspects of an ongoing discourse because they are still processing an initial shocking or unexpected piece of information.

Impact: The Attentional Blink can cause us to overlook important details and limit our understanding. It's as if our brain has a small buffer and sometimes, well, it overflows.

Related Biases: Change Blindness and Inattentional Blindness could be considered close relatives of the Attentional Blink.

Countermeasures: Mindfulness and training can potentially decrease the effect of this bias. It's like mental yoga for your cognitive functions.

Key Takeaways: The Attentional Blink reminds us that our brain, while wonderful, can also be wonderfully fallible. So, slow down and try not to blink—you might just miss something important.

References:

Raymond, J. E., Shapiro, K. L., & Arnell, K. M. (1992). Temporary suppression of visual processing in an RSVP task: An attentional blink?. Journal of Experimental Psychology: Human Perception and Performance, 18(3), 849–860.

Dux, P. E., & Marois, R. (2009). The attentional blink: A review of data and theory. Attention, Perception, & Psychophysics, 71(8), 1683–1700.

Authority Bias

Definition: Authority Bias is our human tendency to value and follow the opinions or actions of perceived authority figures, sometimes to a fault.

Description: This bias stems from the logical heuristic that authorities usually possess more accurate and relevant information. However, the shortcut can backfire when we overgeneralize, accepting authorities as infallible and ignoring our own judgment.

Examples: In religion, people might believe a doctrine because a respected religious leader endorses it, overlooking personal doubts or inconsistencies. "Father McVirtuous said it, so it must be true!"

In politics, we see Authority Bias play out when voters blindly follow political leaders' views or party lines without questioning or fact-checking. "If President Highhorse says so, who am I to argue?"

As for social issues, Authority Bias can lead people to accept problematic cultural norms or societal standards because influential figures or institutions support them. "Celebrity X is wearing animal fur, it must be ethical, right?"

Impact: Authority Bias can lead to uncritical acceptance of flawed ideas, misinformation, and harmful behavior. This bias can also perpetuate social inequalities and allow abuses of power.

Related Biases: The Halo Effect, Expert Bias, and Bandwagon Effect are closely related biases.

Countermeasures: Always approach authority with a healthy dose of skepticism. Foster critical thinking, fact-checking, and diversity of perspectives. Keep in mind that being an authority in one area doesn't grant infallibility in all.

Key Takeaways: Authority Bias reminds us that a suit, a title, or a large following doesn't guarantee correctness. Always think for yourself, even when confronted with a badge, a crown, or a blue tick on Twitter.

References:

Cialdini, R. B. (2001). Influence: Science and practice (4th ed.). Boston: Allyn & Bacon.

Milgram, S. (1963). Behavioral study of obedience. The Journal of Abnormal and Social Psychology, 67(4), 371–378.

Automation Bias

Definition: Automation Bias refers to our tendency to favor decisions or recommendations generated by automated systems over those made by humans—even when they are wrong.

Description: As technology invades every facet of our lives, we tend to assign an unearned level of infallibility to machines. This bias is rooted in our belief that automation is generally more reliable, more precise, and less prone to error than humans.

Examples: In religion, think of the plethora of horoscope apps or AI-generated "personalized" spiritual advice. "My AI-powered tarot app predicted a day full of joy. Let's ignore that grouchy feeling in the morning."

In politics, people might uncritically accept AI-curated news feeds, allowing algorithms to shape their political views. "The algorithm feeds me only left-wing articles—it must know what's best for me."

On social issues, automated systems like predictive policing algorithms might be used unquestioningly, despite potential biases in the data. "The AI predicts crime will occur here—let's ignore socio-economic factors and historical bias in data collection."

Impact: Automation Bias can lead to over-reliance on technology, loss of critical thinking, and ignorance of human intuition or expertise. It can also amplify existing biases in AI systems, potentially perpetuating societal injustices.

Related Biases: The Trust Bias and Confirmation Bias are notable cousins of Automation Bias.

Countermeasures: Strive for a balanced relationship with technology. Do not discard human intuition, expertise, or oversight. Make sure to question automated outcomes and remain vigilant for biases in the system.

Key Takeaways: Automation is a tool, not a god. Machines make mistakes—sometimes even more spectacularly than we humans do. Don't forget to wear your skeptical glasses even when dealing with high-tech gizmos.

References:

Goddard, K., Roudsari, A., & Wyatt, J. C. (2014). Automation bias: empirical results assessing influencing factors. International Journal of Medical Informatics, 83(5), 368–375.

Parasuraman, R., & Manzey, D. H. (2010). Complacency and bias in human use of automation: An attentional integration. Human Factors, 52(3), 381–410.

Availability Cascade

Definition: The Availability Cascade is a self-reinforcing cycle in which a claim gains more acceptance as it is repeated and shared, often regardless of its factual validity.

Description: This bias dances to the tune of "repeat a lie often enough and it becomes the truth." As more people hear and spread a notion, its perceived credibility increases, and it becomes more "available" in people's minds. The old saying about "no smoke without fire" is a delightful example of this bias in action.

Examples: In religion, a misconception that certain religious groups are more violent than others can become widely accepted if repeated often enough in the media or social circles.

In politics, if a politician repeats a catchy, yet factually incorrect, slogan enough times, it may become accepted as truth by a large portion of the populace. "No, we can't just print more money to pay off the national debt."

Regarding social issues, unfounded rumors about a marginalized group can become accepted 'common knowledge' if they're repeated enough times. "Contrary to popular belief, not all millennials eat avocado toast."

Impact: The Availability Cascade can shape public opinion, affect policy decisions, and even fuel mass hysteria. It can lead to the acceptance of misconceptions and stereotypes, which can have damaging consequences on individual and societal levels.

Related Biases: This bias is a close kin to the Bandwagon Effect, the Illusory Truth Effect, and Confirmation Bias.

Countermeasures: Make friends with skepticism. Verify information from multiple sources before accepting or spreading it. Also, remember: just because something is widely accepted, doesn't mean it's true.

Key Takeaways: In a world of 'fake news' and 'alternative facts', a healthy dose of skepticism and fact-checking is your best defense against the Availability Cascade. Remember, if something seems too good (or too bad) to be true, it probably is.

References:

Kuran, T., & Sunstein, C. R. (1999). Availability cascades and risk regulation. Stanford Law Review, 51(4), 683–768.

Sunstein, C. R. (2018). #Republic: Divided democracy in the age of social media. Princeton University Press.

Availability Heuristic

Definition: The Availability Heuristic is a mental shortcut that relies on immediate examples coming to mind when evaluating a specific topic, concept, method, or decision.

Description: Human brains, like the best lazy rivers, prefer the path of least resistance. When we need to make a decision or judgement, we often use what comes to mind quickly. The more easily a piece of information is recalled, the more "true" or prevalent we assume it to be. Unfortunately, our minds are less like a well-indexed library and more like a toddler's toy box - the most accessed pieces often end up on top.

Examples:

Religion: "My friend, Bob, who is very religious, is a kind person. Hence, all religious people must be kind."

Politics: "I've seen several news stories about corrupt politicians. Clearly, all politicians must be corrupt."

Social Issues: "I recently saw a documentary on homelessness in big cities. Hence, homelessness must be solely an urban issue."

Impact: The Availability Heuristic can lead to overestimations of frequency or probability, based on how easily examples are remembered. This can lead to stereotypes, poor decision making, and misjudgments of risk.

Related Biases: This bias shares a few genes with the Recency Effect, the Salience Bias, and the Confirmation Bias.

Countermeasures: Slow down your thinking process and seek out more comprehensive, statistically sound information before making a judgment or decision.

Key Takeaways: Just because something springs readily to mind doesn't mean it's representative or true. When making judgments or decisions, strive for a well-rounded perspective, not just the low-hanging cognitive fruit.

References:

Tversky, A., & Kahneman, D. (1973). Availability: A heuristic for judging frequency and probability. Cognitive psychology, 5(2), 207-232.

Schwarz, N., Bless, H., Strack, F., Klumpp, G., Rittenauer-Schatka, H., & Simons, A. (1991). Ease of retrieval as information: Another look at the availability heuristic. Journal of Personality and Social Psychology, 61(2), 195–202.

Backfire Effect

Definition: The Backfire Effect is the phenomenon in which individuals strengthen their beliefs when presented with contradictory evidence, effectively backfiring against efforts to correct misinformation.

Description: The Backfire Effect is like that old rusted-out car in the yard that only runs faster the more you kick it. Present people with evidence contrary to their deeply held beliefs, and rather than reassessing, they double down. It's a fascinating display of stubbornness that underscores the complexities of cognitive processing and the interplay between emotions and beliefs.

Examples:

Religion: "New archaeological evidence contradicts a core belief of my religion? Well, clearly, the archaeologists are biased against my faith."

Politics: "Fact-checkers say my favorite politician lied? They must be part of the opposition party."

Social Issues: "A study says our city's crime rate is down, yet I feel it's up. Obviously, the researchers are cooking the numbers."

Impact: The Backfire Effect can hinder critical thinking and perpetuate misconceptions, leading to polarized debates and even policy decisions based on false information.

Related Biases: Confirmation Bias and Belief Perseverance might as well be the Backfire Effect's cousins at a family barbecue - each reinforcing the other.

Countermeasures: Foster open-mindedness and curiosity, encourage critical thinking, and emphasize the importance of evidence-based beliefs. Remember, it's better to grasp the nettle of truth than hug the cactus of falsehood.

Key Takeaways: If evidence against your belief makes you believe it more strongly, you might be doing thinking wrong. Reassess your beliefs with a critical eye and be willing to change your mind when confronted with strong evidence.

References:

Nyhan, B., & Reifler, J. (2010). When corrections fail: The persistence of political misperceptions. Political Behavior, 32(2), 303-330.

Lewandowsky, S., Ecker, U. K., & Cook, J. (2017). Beyond Misinformation: Understanding and Coping with the "Post-Truth" Era. Journal of Applied Research in Memory and Cognition, 6(4), 353-369.

Backward Induction

Definition: Backward Induction is a decision-making process where one starts from the end of a problem and works backwards to the beginning, much like reading a mystery novel from back to front to see who done it right off the bat.

Description: In theory, Backward Induction is a neat way to solve multi-stage decision problems. But humans are not always rational creatures. Rather than logically calculating from the end to the beginning, we're often swayed by emotions, biases, and the sheer novelty of reading a book backwards.

Examples:

Religion: "If the world ends in Armageddon, then the events leading up to it are... Wait, this isn't making my faith more reassuring."

Politics: "If we want the bill passed in 2 years, we need to secure votes, which means convincing our colleagues, which means...oh dear, politics is hard."

Social Issues: "If we want to eradicate poverty by 2040, then in 2039 we need...Oh no, this makes the problem seem even more insurmountable."

Impact: Backward Induction can lead to suboptimal or even harmful decisions if not properly used. It can also create anxiety when the end of the problem seems unreachable from the current state.

Related Biases: This isn't exactly a bias, more of a strategy prone to the influence of biases. But it shares the stage with the Hindsight Bias, Overconfidence Bias, and a whole raft of cognitive biases that love to stick their noses into your decision-making process.

Countermeasures: Use it with caution. Break down large problems into manageable tasks. Pair it with other decision-making tools and remain flexible. Remember, it's not about reading the book backwards; it's about understanding the story.

Key Takeaways: Backward Induction can be a useful tool, but it's not the be-all and end-all of decision making. Be aware of your cognitive biases, and don't be afraid to approach problems from different angles.

References:

Binmore, K. (1987). Modeling rational players: Part I. Economics and philosophy, 3(2), 179-214.

Camerer, C. (2003). Behavioral Game Theory: Experiments in Strategic Interaction. Princeton University Press.

Backward Reasoning

Definition: Backward Reasoning is the method of solving a problem by starting with the desired outcome and reasoning backwards to determine the steps needed to reach that outcome. Imagine it's like a reverse-engineered thought process or your brain doing the moonwalk.

Description: Backward Reasoning, though not technically a cognitive bias, can be prone to the influence of biases and other cognitive distortions. This approach relies heavily on the ability to visualize a goal and reason out the necessary prerequisites. However, because we're not in a well-structured board game, reality has a habit of introducing unpredictable factors that can render initial reasoning faulty.

Examples:

Religion: "If my aim is to reach heaven, then I need to follow all the religious commandments, which means...Hey, this could be a helpful tool for religious piety, or the path to fanaticism."

Politics: "If we want to win the next election, we need popular policies, which means we need to...Oh no, I'm starting to sound like a politician."

Social Issues: "If we want to reduce unemployment, we need to stimulate the economy, which means we need...I feel like I just solved the economic crisis, but did I really?"

Impact: Backward Reasoning can lead to strategic thinking and detailed planning, but it can also result in oversimplification of complex problems, blind spots, and an overreliance on ideal scenarios.

Related Biases: This reasoning strategy can be affected by the Confirmation Bias, Optimism Bias, and Planning Fallacy, among others, as people might overlook unfavorable steps or overestimate their ability to execute the plan.

Countermeasures: Counter the influence of biases by encouraging diverse perspectives, testing assumptions, and planning for contingencies. Like with a game of chess, it's not just about the final checkmate, but also about navigating the dynamic landscape of the board.

Key Takeaways: Backward Reasoning is a powerful tool but is not immune to biases and distortions. It is best used in conjunction with other tools and approaches, and with a healthy degree of critical thinking.

References:

Rieskamp, J., Busemeyer, J. R., & Mellers, B. A. (2006). Extending the bounds of rationality: Evidence and theories of preferential choice. Journal of Economic Literature, 44(3), 631-661.

Hastie, R., & Dawes, R. M. (2001). Rational choice in an uncertain world: The psychology of judgment and decision making. Sage Publications.

Bandwagon Effect

Definition: The Bandwagon Effect is a cognitive bias that influences an individual to believe or do things because many other people do or believe the same. It's the psychological phenomenon behind the phrase "jump on the bandwagon."

Description: Rooted in social proof, a form of normative social influence, the Bandwagon Effect nudges individuals to conform to the majority, regardless of the sensibility or morality of the action or belief. It reflects our intrinsic desire to fit in, be part of a group, and feel accepted. This psychological quirk could make the black sheep of us all seem like the fluffiest of lambs, given enough peer pressure.

Examples: The Bandwagon Effect runs amok in many aspects of life. In politics, candidates' popularity often swells if they're perceived as the likely winner - nothing attracts a crowd like a crowd, right? In social media, a post might garner more likes purely because it already has a considerable number. In religion, people may adopt beliefs espoused by the majority in their community, even if these contradict personal instincts or logic.

Impact: It can significantly influence decision-making and perception, leading to herd mentality, and possibly, unwise decisions. It also has implications for mass behavior and the propagation of societal norms, even those that might be unconstructive or harmful.

Related Biases: Groupthink, Herd Behavior, and Conformity Bias share similar features with the Bandwagon

Effect, emphasizing collective consensus over individual judgment.

Countermeasures: Critical thinking and individual analysis are antidotes to the Bandwagon Effect. Questioning why you're adopting a belief or taking an action and seeking diverse information sources can counteract the pull of the crowd.

Key Takeaways: The Bandwagon Effect is a potent force shaping our beliefs and behaviors. Recognizing it can help us make more informed decisions and resist the allure of mass approval. We should remember that popular doesn't always mean right.

References:

Deutsch, M., & Gerard, H. B. (1955). A study of normative and informational social influences upon individual judgment. The Journal of Abnormal and Social Psychology, 51(3), 629–636. doi: 10.1037/h0046408.

Leibenstein, H. (1950). Bandwagon, Snob, and Veblen Effects in the Theory of Consumers' Demand. The Quarterly Journal of Economics, 64(2), 183–207. doi: 10.2307/1882692.

Base Rate Neglect

Definition: Base Rate Neglect, or Base Rate Fallacy, is a cognitive bias where individuals disregard or underestimate the fundamental 'base rate' information in favor of specific, albeit less consequential, details when making decisions. It's the mental equivalent of forgetting the forest for a single, charismatic tree.

Description: It's a common cognitive hiccup in our probability assessments. This bias has us, often incorrectly, discounting statistical information (the base rate) when it's presented alongside more vivid, individual, or anecdotal information. The human brain, it seems, is a sucker for a good story, even if that story tells us to ignore the boring yet relevant statistics.

Examples: Imagine hearing about a politician embroiled in scandal. The base rate data might suggest that corruption is relatively rare among politicians. However, the specific, salacious details of the scandal, broadcasted on the 24/7 news cycle, could lead many to believe such misbehavior is more prevalent than it is. Or, in the context of religion, a miraculous healing story might lead people to overestimate the effectiveness of faith healing, despite clinical data suggesting otherwise.

Impact: Base Rate Neglect can lead to misinformed beliefs and poor decision-making across multiple spheres, including personal health, public policy, and societal perceptions. This bias can also contribute to stereotyping and discrimination when we overemphasize certain characteristics or behaviors among a group.

Related Biases: The Representative Heuristic, a tendency to judge probabilities based on how much something resembles a specific category, can often co-exist with Base Rate Neglect.

Countermeasures: Improving statistical literacy and emphasizing the importance of base rates can help. Additionally, fostering a habit of asking "what's the bigger picture?" or "what's the underlying data?" can lead to more balanced decision-making.

Key Takeaways: Don't get swept up in the specific or flashy details at the expense of the broader, more objective data. Base Rate Neglect is a reminder to consider all relevant information, especially the more statistical and perhaps less glamorous facts, before making decisions or forming beliefs.

References:

Tversky, A., & Kahneman, D. (1974). Judgment under Uncertainty: Heuristics and Biases. Science, 185(4157), 1124-1131. doi:10.1126/science.185.4157.1124.

Bar-Hillel, M. (1980). The base-rate fallacy in probability judgments. Acta Psychologica, 44(3), 211-233. doi:10.1016/0001-6918(80)90046-3.

Belief Bias

Definition: Belief Bias is the tendency to evaluate reasoning by the believability of its conclusion rather than how logically sound it is. Essentially, it's when our brains cheer, "That's what I've been saying all along!" and conveniently ignore the potential logical flaws in our argument.

Description: This bias is a cognitive speed bump that often prevents us from thinking critically. Our pre-existing beliefs, opinions, and values take precedence over logic in our reasoning process. When we stumble upon information or arguments that align with our current beliefs, we tend to accept them as true, often without proper scrutiny. Conversely, we're more likely to be suspicious of and discard information that contradicts our beliefs, even if it's logically sound.

Examples: A classic example can be found in politics. Suppose a person strongly supports a particular political party. They're likely to accept any reasoning or argument that paints their party in a good light, and dismiss those that criticize it, regardless of the merits of the arguments. Similarly, a deeply religious individual might dismiss scientifically sound arguments that challenge their beliefs.

Impact: The Belief Bias can lead to ill-informed decisions, polarized views, and inability to have productive discussions. It can contribute to the divide in society on many critical issues, from climate change to public health.

Related Biases: Confirmation Bias, the tendency to seek, interpret, and remember information that confirms our pre-existing beliefs, is a close cousin of Belief Bias.

Countermeasures: Emphasizing critical thinking skills, promoting empathy and understanding of other viewpoints, and encouraging an open-minded approach to information can help counteract this bias.

Key Takeaways: Belief Bias reminds us that our pre-existing beliefs don't always make the best judges. To make informed decisions, it's important to separate what we want to believe from what is logically sound.

References:

Evans, J. B. T., Barston, J. L., & Pollard, P. (1983). On the conflict between logic and belief in syllogistic reasoning. Memory & Cognition, 11(3), 295–306. https://doi.org/10.3758/BF03196976

Oakhill, J., Johnson-Laird, P. N., & Garnham, A. (1989). Believability and syllogistic reasoning. Cognition, 31(2), 117–140. https://doi.org/10.1016/0010-0277(89)90014-3

Belief Perseverance

Definition: Belief Perseverance is the cognitive bias that leads people to hold on to their beliefs even when they're presented with evidence that directly contradicts them. It's like our brain is singing, "I will survive!" while stubbornly clinging to its treasured beliefs.

Description: This bias often occurs when we invest significant time, energy, or emotion into a belief. Even in the face of clear disconfirming evidence, we manage to perform impressive cognitive gymnastics to maintain our belief. This can involve discrediting the evidence, its source, or concocting new justifications for the belief. It's our brain's attempt at self-preservation, defending our identity, and maintaining consistency in our world view.

Examples: If an individual firmly believes in a political conspiracy theory, even overwhelming factual evidence contradicting the theory may not change their belief. They might dismiss the evidence as part of the conspiracy, or accuse the source of the evidence as biased or unreliable.

Impact: Belief Perseverance can lead to dangerous misinformation, polarization in society, and resistance to change. It can prevent progress in important areas like science, social justice, and policy-making.

Related Biases: Confirmation Bias and Backfire Effect are related biases, with all revolving around our tendency to protect and defend our existing beliefs.

Countermeasures: Encouraging skepticism, fostering an open-minded attitude, and promoting a culture of changing one's mind in light of new evidence can help mitigate this bias.

Key Takeaways: Belief Perseverance shows how stubborn our minds can be. While it's comfortable to stick with what we know, growth and truth often lie in challenging our beliefs and being willing to update them.

References:

Ross, L., Lepper, M. R., & Hubbard, M. (1975). Perseverance in self-perception and social perception: biased attributional processes in the debriefing paradigm. Journal of Personality and Social Psychology, 32(5), 880–892. https://doi.org/10.1037/0022-3514.32.5.880

Anderson, C. A., Lepper, M. R., & Ross, L. (1980). Perseverance of social theories: The role of explanation in the persistence of discredited information. Journal of Personality and Social Psychology, 39(6), 1037–1049. https://doi.org/10.1037/0022-3514.39.6.1037

Ben Franklin Effect

Definition: The Ben Franklin Effect is a psychological phenomenon where we tend to like someone more after we do them a favor. It seems our brains got the memo that "it's better to give than to receive," but missed the part about not expecting anything in return.

Description: This cognitive bias refers to the surprising tendency of people to like others more once they've done a favor for them, even if they initially disliked the person. This effect is an example of cognitive dissonance—our brains find it difficult to reconcile the fact that we did something nice for someone we don't like, so we decide we must like them after all. We humans are funny creatures.

Examples: Suppose a politician does not personally like a rival from an opposing party. But if they assist the rival in a legislative effort, they might find their initial dislike softening.

Impact: The Ben Franklin Effect can have a significant impact on social and professional relationships, including diplomacy and politics. Understanding it can help improve personal interactions and facilitate smoother negotiations in polarized environments.

Related Biases: This bias is related to the cognitive dissonance theory and the self-perception theory, both dealing with our mind's knack for resolving internal conflict to preserve a consistent self-image.

Countermeasures: Being aware of the Ben Franklin Effect can help individuals understand their emotional changes after doing favors and may guide them to separate actions from personal feelings.

Key Takeaways: The Ben Franklin Effect is a unique quirk of human psychology that shows our mind's need for consistency. The next time you find yourself warming up to someone for whom you've done a favor, you might just be experiencing this effect.

References:

Jecker, J., & Landy, D. (1969). Liking a person as a function of doing him a favour. Human Relations, 22(4), 371–378. https://doi.org/10.1177/001872676902200407

Aronson, E. (1999). Dissonance, Hypocrisy, and the Self-Concept. In E. Harmon-Jones & J. Mills (Eds.), Cognitive Dissonance: Progress on a Pivotal Theory in Social Psychology. American Psychological Association. https://doi.org/10.1037/10318-000

Bias Blind Spot

Definition: The Bias Blind Spot is the cognitive bias of recognizing the impact of biases on the judgment of others, while failing to see the impact of biases on one's own judgment. It's like believing you're the only driver on the road who knows what they're doing.

Description: This psychological phenomenon refers to our impressive ability to spot biases in others, while being as blind as a bat to our own. The Bias Blind Spot demonstrates our fabulous knack for maintaining a self-image that is infallible, while everyone else is about as perfect as a Picasso painting... if it were drawn by a four-year-old.

Examples: Imagine a politician accusing a rival of being biased in their views on immigration because they're from a border state, while failing to recognize that their own views could be biased by their upbringing in a land-locked state with low immigration.

Impact: The Bias Blind Spot can significantly influence judgment and decision-making, often leading to overconfidence in personal beliefs and underestimation of influences of bias. This can impact discussions around contentious issues in politics, religion, and social matters.

Related Biases: The Bias Blind Spot is related to the Dunning-Kruger effect, where individuals with low ability at a task overestimate their ability.

Countermeasures: Reflecting on our own potential for bias and seeking external perspectives can help mitigate the influence of the Bias Blind Spot. Embracing humility wouldn't be such a bad idea either.

Key Takeaways: The Bias Blind Spot reminds us that we're not as perfect as we think we are and that it might be a good idea to put on our bias-detecting goggles when examining our own judgments.

References:

Pronin, E., Lin, D. Y., & Ross, L. (2002). The bias blind spot: Perceptions of bias in self versus others. Personality and Social Psychology Bulletin, 28(3), 369–381. https://doi.org/10.1177/0146167202286008

Kruger, J., & Dunning, D. (1999). Unskilled and unaware of it: how difficulties in recognizing one's own incompetence lead to inflated self-assessments. Journal of personality and social psychology, 77(6), 1121–1134.

Boomerang Effect

Definition: The Boomerang Effect is a cognitive bias where attempts to persuade people to change their beliefs or behavior can backfire and strengthen their existing attitudes instead. It's like telling a toddler not to touch the cookie jar, thereby guaranteeing the imminent disappearance of all cookies.

Description: This amusingly stubborn phenomenon occurs when people respond to persuasive messages by adopting an opposing position, rather like a boomerang returning to hit you right in the common sense. People often react this way to protect their freedom of thought and avoid feeling manipulated, making it the cognitive equivalent of sticking your fingers in your ears and yelling "la la la, I can't hear you."

Examples: For instance, a political campaign might release an ad attacking a candidate's opponent. Instead of winning votes, the ad could end up increasing support for the opponent, especially if the public perceives the ad as a dirty trick.

Impact: The Boomerang Effect can lead to the reinforcement of harmful beliefs or behaviors, especially in contentious areas like religion, politics, and social issues. It's the unwanted guest who not only crashes your party but also drinks all your best wine.

Related Biases: This is related to the Backfire Effect, where exposure to facts contradicting one's beliefs strengthens those beliefs rather than changing them.

Countermeasures: To mitigate the Boomerang Effect, communicators should consider their audience's beliefs and values before sending a persuasive message. Also, try being subtle and more diplomatic in approach, like a cat stalking its prey, rather than a bull in a china shop.

Key Takeaways: The Boomerang Effect serves as a stark reminder that not all attempts to change minds are successful, and sometimes, they can even make things worse.

References:

Silvia, P. J. (2006). Reactance and the dynamics of disagreement: Multiple paths from threatened freedom to resistance to persuasion. European Journal of Social Psychology, 36(5), 673–685. https://doi.org/10.1002/ejsp.309

Nyhan, B., & Reifler, J. (2010). When Corrections Fail: The Persistence of Political Misperceptions. Political Behavior, 32(2), 303-330. https://doi.org/10.1007/s11109-010-9112-2

Bystander Effect

Definition: The Bystander Effect refers to the psychological phenomenon in which individuals are less likely to offer help to a victim when others are present. It's the cognitive equivalent of "someone else will deal with it."

Description: Rooted in the diffusion of responsibility, the Bystander Effect essentially operates on the principle that with greater numbers comes less personal accountability. It's like the mental version of the phrase, "Too many cooks spoil the broth," except the broth is a person in need of help, and the cooks are too busy expecting other cooks to act.

Examples: Imagine a politician is caught up in a scandal, and everyone in the party is aware of it. Instead of stepping forward, each individual assumes someone else will address it. The result? A whole lot of inaction, and a scandal that continues to sizzle.

Impact: The Bystander Effect can have grave consequences, from minor incidents escalating unaddressed to severe social issues, such as hate crimes or discrimination, going unchecked. It's like a bystander ignoring a leaking faucet, only to be shocked when the house floods.

Related Biases: The Bystander Effect is related to the Diffusion of Responsibility, where the responsibility to act is divided among group members, leading each person to feel less obligated to act.

Countermeasures: Education and awareness about the Bystander Effect can motivate individuals to break the chain of inaction. Also, specific calls to action can help, as direct responsibility tends to buckle the knees of the Bystander Effect.

Key Takeaways: The Bystander Effect is a psychological quirk reminding us that more people do not necessarily equate to more action. Recognizing the bias can prompt us to become the "someone" in "someone should do something."

References:

Darley, J. M., & Latané, B. (1968). Bystander intervention in emergencies: Diffusion of responsibility. Journal of Personality and Social Psychology, 8(4), 377–383. https://doi.org/10.1037/h0025589

Fischer, P., Krueger, J. I., Greitemeyer, T., Vogrincic, C., Kastenmüller, A., Frey, D., ... & Kainbacher, M. (2011). The bystander-effect: a meta-analytic review on bystander intervention in dangerous and non-dangerous emergencies. Psychological Bulletin, 137(4), 517–537. https://doi.org/10.1037/a0023304

Cheerleader Effect

Definition: The Cheerleader Effect, also known as the group attractiveness effect, is a cognitive bias where people are perceived as more attractive when seen in a group rather than individually. It's like saying, "Together we're a 10, individually...well, that's subjective."

Description: This curious bias has its roots in the way our brains process visual information. When viewing a group, our brains tend to "average out" the features of individual members, smoothing over the less attractive aspects and highlighting the more attractive ones. It's kind of like the cognitive equivalent of squinting to make something look better.

Examples: Ever wondered why political parties like to display their members en masse? It's not just for the group photo. The Cheerleader Effect can make them appear more appealing (if we're talking strictly visual perception, of course).

Impact: On a superficial level, the Cheerleader Effect could influence social dynamics, even electoral outcomes. For example, a party or candidate might seem more appealing when seen as part of a group. On a deeper level, it highlights how our perception can be easily skewed by context.

Related Biases: The Halo Effect is a related bias, where the perception of one positive attribute (like attractiveness) influences the perception of other attributes.

Countermeasures: Recognizing that our brains tend to "average out" features in a group can help us take a second look and evaluate individuals on their own merits. It's like seeing past the group discount to judge each item's value separately.

Key Takeaways: The Cheerleader Effect shows us that context matters, and our perception of attractiveness can be influenced by group dynamics. It reminds us to look beyond the immediate context and consider individuals independently.

References:

Walker, D., & Vul, E. (2014). Hierarchical Encoding Makes Individuals in a Group Seem More Attractive. Psychological Science, 25(1), 230–235. https://doi.org/10.1177/0956797613497969

Choice-Supportive Bias

Definition: Choice-supportive bias is a cognitive bias where individuals remember their past choices as better than they actually were. It's like applying rose-tinted glasses to your past decisions. Suddenly, that second-hand car that broke down every other day was a 'classic' with 'character'.

Description: This bias is a result of our desire for consistency and our natural aversion to cognitive dissonance (that uncomfortable feeling when our actions don't match our beliefs). We'd rather revise our memories to feel better about our decisions, rather than admit we made a subpar choice. Memory, it turns out, is not only selective but also rather diplomatic.

Examples: Ever noticed how some people swear by their political party even when confronted with its failures? Or how about that friend who insists their terrible ex was "misunderstood"? That's choice-supportive bias in action. They've simply chosen to remember their past decision (to support a party or date someone) in a positive light.

Impact: This bias can limit personal growth and cause repeated poor decision-making. Imagine sticking to a bad investment or a flawed strategy simply because admitting failure seems too uncomfortable.

Related Biases: Confirmation bias, where individuals focus on information that confirms their pre-existing beliefs, and hindsight bias, where people believe past events were more predictable than they were.

Countermeasures: Seek out and consider disconfirming evidence about your past choices. Don't let the discomfort of cognitive dissonance stop you from making more accurate assessments. Sometimes, the best way to make better decisions in the future is to recognize and learn from past mistakes.

Key Takeaways: Choice-supportive bias shows us that our memories are not always trustworthy, particularly when it comes to justifying our past decisions. Recognizing this bias can help us to better assess our decisions and learn from past mistakes.

References:

Mather, M., Shafir, E., & Johnson, M. K. (2000). Misrememberance of options past: source monitoring and choice. Psychological science, 11(2), 132–138. https://doi.org/10.1111/1467-9280.00228

Chameleon Effect

Definition: The chameleon effect refers to our unconscious mimicry of others' behaviors, postures, and mannerisms. It's the reason why you might find yourself leaning in when someone else leans in during a conversation, or why you start using "y'all" after watching too many Westerns.

Description: This effect is an aspect of social cognition that helps us form connections and fit into social situations. Evolutionarily, it's been a handy tool for surviving tribal societies, and let's be honest, high school. It's driven by our subconscious mind - like a "monkey see, monkey do" protocol hardwired into our brains.

Examples: Ever found yourself mimicking the accent of the person you're talking to, much to your embarrassment? Or realized that you've started using the same catchphrases as your favorite TV show character? Welcome to the chameleon effect in action.

Impact: While the chameleon effect can foster rapport and social harmony, it can also lead to groupthink or conformity, limiting diversity of thought. Additionally, it might make you adopt habits, attitudes, or actions that may not align with your own beliefs or principles, if you're not aware of it.

Related Biases: Conformity bias, which refers to our tendency to behave similarly to others in a group, even when it contradicts our own judgment.

Countermeasures: Cultivate self-awareness. Recognizing when you're mirroring others can help you determine if it's appropriate or beneficial. Maintain a clear sense of your own values, principles, and identity to help guard against unconscious conformity.

Key Takeaways: The chameleon effect is a subconscious process that helps us fit into social situations by mimicking others' behavior. While it can help build rapport, it can also lead to conformity and loss of personal identity. By cultivating self-awareness, you can better manage this effect.

References:

Chartrand, T. L., & Bargh, J. A. (1999). The chameleon effect: The perception–behavior link and social interaction. Journal of personality and social psychology, 76(6), 893. https://doi.org/10.1037/0022-3514.76.6.893

Clustering Illusion

Definition: The Clustering Illusion is a cognitive bias that leads individuals to see patterns in random sequences or data where none actually exist.

Description: Our brains love a good story - something that makes sense and ties neatly together. The Clustering Illusion is an example of this, where we tend to find order and meaningful patterns in randomness. This is because humans are inherently wired to perceive patterns as part of our survival instinct.

Examples: For instance, imagine observing a series of coin flips. If you get a few heads in a row, you might start to think there's a pattern forming, but really, it's just randomness at play. On a broader scale, think about how we read into political polling data or religious texts, often perceiving patterns or trends that support our existing beliefs.

Impact: This bias can lead to poor decision-making in various fields, from finance (interpreting stock market trends) to politics (drawing conclusions from polling data) and religion (seeing patterns in religious texts that validate personal beliefs).

Related Biases: Apophenia (the tendency to mistakenly perceive connections and meaning between unrelated things) and Pareidolia (seeing specific images in random or ambiguous visual patterns).

Countermeasures: Understanding statistics and probabilities can help combat this bias. Also, consider the possibility that what you're observing might be random rather than patterned. Critical thinking and skepticism are your best defense mechanisms here.

Key Takeaways: The Clustering Illusion is a mental shortcut that leads us to perceive patterns in randomness. It's essential to recognize that this is a product of our mind's desire for order, and not always indicative of true patterns or trends.

References:

Falk, R., & Konold, C. (1997). Making sense of randomness: Implicit encoding as a basis for judgment. Psychological Review, 104(2), 301. doi:10.1037/0033-295x.104.2.301

Tversky, A., & Kahneman, D. (1971). Belief in the law of small numbers. Psychological Bulletin, 76(2), 105. doi:10.1037/h0031322

Cobra Effect

Definition: The Cobra Effect refers to a situation when an attempted solution to a problem makes the problem worse, due to unforeseen or unintended consequences.

Description: Named after an anecdote from the British colonial rule in India, the Cobra Effect illustrates how our best intentions can sometimes come back and bite us in the rear end, quite like a disgruntled cobra. This is largely due to a failure to account for the fact that people respond to incentives, and not always in the ways we'd expect.

Examples: Consider an attempt to reduce fake news circulation by imposing heavy fines on platforms where it's shared. Rather than deterring the spread, it might encourage the development of more covert methods of distribution. Politically, attempts to impose strict regulations can sometimes incentivize black markets or other undesirable activities. Religiously, a ban on a particular book might only increase its allure, leading to increased demand.

Impact: The Cobra Effect can undermine policy-making, business strategies, and social initiatives. It reinforces the need for careful and comprehensive analysis before implementing solutions, especially in complex systems like politics, economics, and social issues.

Related Biases: Unintended consequences, Law of Unforeseen Consequences, and perverse incentives.

Countermeasures: Rigorous testing, phased rollouts, and continuous feedback can help detect and mitigate the Cobra Effect. Also, considering a wide range of scenarios and consulting diverse perspectives can help in anticipating possible reactions.

Key Takeaways: The Cobra Effect reminds us that solving problems isn't always straightforward. When dealing with complex systems, anticipate possible reactions and be prepared for your solution to potentially exacerbate the issue.

References:

Geckil, I. K., & Nair, R. D. (2010). Perverse incentives: The cobra effect. CFA Institute Magazine, 21(2), 41-45.

Tenner, E. (1996). Why Things Bite Back: Technology and the Revenge of Unintended Consequences. New York: Knopf.

Confirmation Bias

Definition: Confirmation bias is the psychological tendency to seek, interpret, and remember information in a way that confirms one's preexisting beliefs or hypotheses, while giving disproportionately less consideration to alternative possibilities.

Description: Like a picky eater at a buffet, confirmation bias has us loading up our mental plates with comforting, agreeable facts while turning up our noses at anything that looks different. It's rooted in our desire to maintain consistency and avoid discomfort, and it can turn the most open-minded of us into stubborn stick-in-the-muds, particularly when our core beliefs are challenged.

Examples: In politics, one might choose to watch a news channel that aligns with their political leaning, disregarding the other side's perspectives. Religiously, believers may focus on parts of religious texts that support their beliefs, while ignoring those that contradict. Socially, a person might only read reviews that praise their favorite movie, while dismissing the negative ones as "people who just didn't get it".

Impact: Confirmation bias can lead to poor decision-making, misunderstanding, and divisiveness. It can skew our perception of reality, hinder learning and growth, and create echo chambers in our social and professional lives.

Related Biases: Belief bias, echo chamber effect, and selective perception.

Countermeasures: Actively seek out information from different perspectives. Practice empathy and open-mindedness. Be aware of your biases and challenge your own beliefs. Encourage debate and diversity of thought in discussions.

Key Takeaways: Confirmation bias acts like a mental comfort blanket, keeping us snug and safe in our own beliefs. While it's cozy under there, it's important to occasionally stick your head out and see what else the world has to offer. You might find that different perspectives aren't as scary as they seem.

References:

Nickerson, R. S. (1998). Confirmation bias: A ubiquitous phenomenon in many guises. Review of General Psychology, 2(2), 175-220.

Lilienfeld, S. O., Ammirati, R., & Landfield, K. (2009). Giving debiasing away: Can psychological research on correcting cognitive errors promote human welfare? Perspectives on Psychological Science, 4(4), 390-398.

Conjunction Fallacy

Definition: The conjunction fallacy is a cognitive bias that occurs when it is assumed that specific conditions are more probable than a single general one. Basically, it's when we think two things happening together is more likely than just one thing happening.

Description: Conjunction fallacy is the mental equivalent of believing two-for-one deals are always better, even if you only needed one thing in the first place. The problem here is simple: less is often more. In probability terms, the likelihood of two events happening together (in "conjunction") can't be higher than the likelihood of either happening separately.

Examples: Let's venture into politics. Suppose someone believes that a politician being both honest and competent is more likely than them just being competent. That's conjunction fallacy, folks. Or in the realm of religion, someone may think it's more probable for a person to be kind-hearted, generous, and a religious believer than just being kind-hearted and generous.

Impact: Conjunction fallacy can lead to flawed decision-making and unrealistic expectations. It can cloud judgment, making us susceptible to persuasion and manipulation, especially in high-stakes areas like politics, religion, or social issues.

Related Biases: The representativeness heuristic, which can often lead to the conjunction fallacy.

Countermeasures: Slow down your thinking process. Evaluate the likelihood of events separately before considering them jointly. Remember that "less is more" also applies to probability.

Key Takeaways: The conjunction fallacy is the cognitive equivalent of putting too many toppings on your pizza until it's no longer a pizza, but a mess. Keep it simple. Probability prefers the minimalist approach.

References:

Tversky, A., & Kahneman, D. (1983). Extensional versus intuitive reasoning: The conjunction fallacy in probability judgment. Psychological Review, 90(4), 293-315.

Hertwig, R., & Gigerenzer, G. (1999). The 'conjunction fallacy' revisited: How intelligent inferences look like reasoning errors. Journal of Behavioral Decision Making, 12(4), 275-305.

Conservatism Bias

Definition: Conservatism bias is a mental tripwire that causes individuals to undervalue new information, leading them to change their beliefs or estimates insufficiently in light of fresh data. It's like sticking with your grandma's chocolate chip cookie recipe even though you just discovered you're allergic to chocolate.

Description: This bias latches onto our natural tendency to resist change and stick with the known. We may underestimate the relevance of new data, or overvalue our prior beliefs, both of which often result in outdated views staying longer than they're welcome. It's like being at a party where the music has changed, but you're still dancing the Macarena.

Examples: In politics, consider a long-time party member who, despite clear evidence of corruption within the party, resists updating their belief about its integrity. Similarly, in religious contexts, an individual may resist changing their religious views even in light of personal experiences or information that contradicts these beliefs.

Impact: Conservatism bias can hamper our ability to adapt and evolve our views. It can solidify stereotypes, bolster outdated political or religious dogmas, and generally stifle our ability to grow and learn. Like expired milk, outdated beliefs don't usually age well.

Related Biases: Confirmation bias, status quo bias, and anchoring bias.

Countermeasures: Deliberately question and review your beliefs in light of new information. Take into account the source and reliability of new data, and consciously adjust your thinking.

Key Takeaways: Don't be that guy stuck doing the Macarena when everyone else is flossing. Be open to new data, adapt your beliefs accordingly, and embrace change. It's not always easy, but your personal growth depends on it.

References:

Rabin, M., & Schrag, J. L. (1999). First Impressions Matter: A Model of Confirmatory Bias. The Quarterly Journal of Economics, 114(1), 37-82.

Edwards, W. (1968). Conservatism in human information processing. In B. Kleinmuntz (Ed.), Formal representation of human judgment (pp. 17–52). New York: Wiley.

Construal Level Theory

Definition: Construal Level Theory (CLT) is a psychological framework suggesting that our distance from an event—be it in time, space, social connections, or hypothetically—changes how we perceive it. You know how a one-week vacation seems like an eternity when it's a year away, but the night before, you can practically smell the beach? That's CLT at work.

Description: CLT proposes that distant events are perceived in a more abstract manner, focusing on the big picture or overall impression ("I'm going to have such an adventure!"). In contrast, near events are seen in a more concrete, detailed-oriented way ("I need to pack sunscreen, finish that report before I leave, and find a cat sitter.").

Examples: In the realm of politics, policies or promises might sound great in abstract when elections are far off, but as election day approaches, voters might start to scrutinize the specifics. Similarly, the distant concept of heaven might be perceived abstractly in religious contexts, while near-term rituals and requirements are viewed more concretely.

Impact: CLT can affect decision-making and perception. We may be more idealistic about distant events but become more pragmatic as they approach. This can lead to overcommitment, under preparation, or even disillusionment when the concrete reality doesn't match the abstract dream.

Related Biases: Optimism bias, planning fallacy, and rosy retrospection.

Countermeasures: Be aware that your perception changes with distance. Try to think concretely about distant events and abstractly about near ones. Also, recognize that your expectations of the future may be overly optimistic due to their abstract nature.

Key Takeaways: When planning for the future, pack some realism with your optimism. Consider the concrete steps needed to achieve your abstract dreams, and be ready to adapt when life throws you a curveball.

References:

Trope, Y., & Liberman, N. (2003). Temporal Construal. Psychological Review, 110(3), 403–421.

Liberman, N., Trope, Y., & Stephan, E. (2007). Psychological Distance. In A. W. Kruglanski & E. T. Higgins (Eds.), Social psychology: Handbook of basic principles (pp. 353–381). New York, NY: Guilford Press.

Contrast Effect

Definition: The contrast effect is a cognitive bias that skews our perception based on previous stimuli or comparisons. Basically, it's why after watching "The Godfather," your cousin Vinny's indie film might seem like a home video from a family picnic.

Description: The contrast effect can subtly or drastically alter our judgments and decisions, depending on what we have recently experienced or evaluated. It's like our minds have a balance scale, but instead of resetting to zero after each weighing, it sticks to the last weight. This causes us to see lighter things as extra light and heavier things as extra heavy.

Examples: In politics, a moderate candidate may seem extreme if preceded by a very conservative or liberal one. In social issues, individuals may judge their own affluence or poverty in sharp contrast to others in their community or media portrayals. In religion, believers might perceive slight deviations from doctrine as severe contrasts.

Impact: The contrast effect can cause us to undervalue good ideas, overvalue bad ones, or make skewed comparisons that affect our decisions, perceptions, and judgments. It's like getting a dulled sense of taste after eating too many spicy tacos—suddenly, a plain taco seems completely flavorless.

Related Biases: Anchoring bias, framing effect, and halo effect.

Countermeasures: Practice awareness of the contrast effect and strive for objective evaluations. When evaluating something, try to isolate it from previous stimuli or comparisons and judge it on its own merits. Just like when tasting wine, cleanse your mental palate before taking a sip of a new idea.

Key Takeaways: Life isn't a buffet where the previous dish flavors the next one. Be aware of the influence of past experiences on your current judgments. As they say, don't let the taco of the past ruin the taco of the present.

References:

Herr, P. M. (1989). Priming price: prior knowledge and context effects. Journal of Consumer Research, 16(1), 67-75.

Mussweiler, T., & Strack, F. (1999). Hypothesis-consistent testing and semantic priming in the anchoring paradigm: A selective accessibility model. Journal of Experimental Social Psychology, 35(2), 136-164.

Control Bias

Definition: Control bias is the tendency to overestimate our ability to control events, which, to be frank, can be as absurd as believing we can dictate the weather just by doing the anti-rain dance.

Description: This bias reflects a fundamental misunderstanding of the cause-and-effect relationship, where we often inflate the level of control we exercise over situations, causing us to view ourselves as the puppeteers rather than the puppets. It's the psychological equivalent of thinking that by shouting at the TV, you can influence the outcome of a sports match. Spoiler alert: You can't.

Examples: In politics, a person may attribute the success of a policy entirely to their favorite politician, ignoring the countless others involved in the process. In religion, someone might believe that their prayers alone can sway the course of events. In social issues, an individual may think they can single-handedly reverse systemic issues, overlooking the many complex factors at play.

Impact: This bias can lead to overconfidence, poor decision-making, and unnecessary risk-taking, like betting your life savings on a roulette spin because you're "feeling lucky." It can also cause undue stress and blame when the inevitable lack of control becomes apparent.

Related Biases: Illusion of control, overconfidence bias, and optimism bias.

Countermeasures: Develop a more accurate understanding of control through self-awareness, humility, and a dash of skepticism. Think critically about your actual influence and the wider context. Remember, you're not the puppet master of the universe, even if you can make a killer shadow puppet.

Key Takeaways: Understanding the limits of our control can save us from a lot of unnecessary disappointment and misguided efforts. It's not always about us. So, sit back, relax, and accept that some things are beyond our control.

References:

Thompson, S. C. (1999). Illusions of control: How we overestimate our personal influence. Current Directions in Psychological Science, 8(6), 187-190.

Langer, E. J. (1975). The illusion of control. Journal of Personality and Social Psychology, 32(2), 311-328.

Courtesy Bias

Definition: Courtesy bias, or as I like to call it, the "you look great in that outfit" syndrome, is the tendency to give a "polite" response rather than an honest one, especially when the truth could be potentially unpleasant.

Description: It's a social nicety that keeps friendships intact, but also distorts reality like a funhouse mirror. With this bias, you may say you enjoyed that three-hour political rally or church sermon to avoid stepping on toes, while you'd rather have spent the time doing anything else.

Examples: In religion, parishioners may express absolute approval of a religious leader's decisions, even if they harbor reservations. In politics, voters might claim they support a policy because it's socially acceptable, even if they disagree with it privately. And in social situations, individuals might voice approval of a popular movement to avoid being ostracized, even though they may have legitimate criticisms.

Impact: Courtesy bias can lead to a distortion of collective opinions and decisions, creating a mirage of consensus where there isn't one. It can be particularly damaging in democratic processes, research, and any situation where accurate feedback is essential.

Related Biases: Social desirability bias, acquiescence bias.

Countermeasures: Encourage open dialogue and assure people that their honest opinions are welcomed and respected. Make sure people understand that disagreement doesn't equal disrespect, and occasionally, being a bit impolite might be what's needed.

Key Takeaways: Courtesy bias is like an overused Instagram filter: it might make things look good, but it doesn't represent reality. Fostering an environment where honesty is valued over mere agreement can help minimize this bias.

References:

Paulhus, D. L. (1991). Measurement and control of response bias. Measures of personality and social psychological attitudes, 17(1), 17-59.

Podsakoff, P. M., MacKenzie, S. B., Lee, J. Y., & Podsakoff, N. P. (2003). Common method biases in behavioral research: a critical review of the literature and recommended remedies. Journal of applied psychology, 88(5), 879.

Curse of Knowledge

Definition: The Curse of Knowledge, ironically, is not a curse that befalls wizards or librarians, but a cognitive bias where an individual with more information fails to consider the perspective of someone less informed.

Description: This is where being a smarty-pants gets tricky. It can cause over-assumption of others' knowledge and leads to poor communication, or as I call it, the "why don't you get it?" syndrome.

Examples: In politics, a policymaker might use jargon-filled language that's comprehensible to her peers but confounding to the general public. In religion, a religious leader might assume his congregation understands complex theological concepts without a proper explanation. And in social situations, a tech-savvy person might frustrate others with unnecessary tech jargon.

Impact: The curse can cause misunderstandings, create barriers in communication, and hamper decision-making processes. Its effect is most noticeable in the fields of education, leadership, and communication where understanding and effective conveyance of ideas is paramount.

Related Biases: False consensus effect, mind projection fallacy.

Countermeasures: To break the curse, try to put yourself in the shoes of the less informed. Simplify your language, ask for feedback to ensure understanding, and embrace the art of explanation.

Key Takeaways: The Curse of Knowledge shows that it's not enough to be knowledgeable. The ability to share that knowledge effectively, sans jargon, is the real magic.

References:

Birch, S. A., & Bloom, P. (2007). The curse of knowledge in reasoning about false beliefs. Psychological Science, 18(5), 382-386.

Camerer, C., Loewenstein, G., & Weber, M. (1989). The curse of knowledge in economic settings: An experimental analysis. Journal of Political Economy, 97(5), 1232-1254.

Decoy Effect

Definition: The Decoy Effect, also known as "asymmetric dominance," isn't a rogue secret agent maneuver, but a cognitive bias where a third option is introduced to make one of the existing choices more attractive.

Description: Picture this - you're torn between two options, like a politician wavering between two policies. Then, in walks a third option that's clearly inferior to one of the existing ones, making that one seem like a supermodel of choice. Congratulations, you've been decoyed! This occurs because humans often rely on relative comparisons rather than absolute evaluations.

Examples: In politics, a decoy candidate could be put forward to draw votes away from the opposition, making the preferred candidate look better. In religion, an extreme interpretation of a belief may be presented to make a less extreme belief seem more moderate and appealing. And in social issues, the presentation of an extreme viewpoint could make a less extreme one appear more reasonable.

Impact: The Decoy Effect can lead us to make decisions that we think are optimal but are really influenced by the context in which choices are presented. It's a crafty tool for manipulation in decision-making, often used in marketing, policy making, and social debates.

Related Biases: Anchoring bias, framing effect, contrast effect.

Countermeasures: Keep a watchful eye for any decoys sneaking around. Evaluate options individually, in absolute terms, not just relative to other options presented.

Key Takeaways: When deciding, don't let the ugly duckling (decoy) make the others look better. Every option should strut down the decision-making runway independently.

References:

Huber, J., Payne, J. W., & Puto, C. (1982). Adding asymmetrically dominated alternatives: Violations of regularity and the similarity hypothesis. Journal of Consumer Research, 9(1), 90-98.

Pettibone, J. C., & Wedell, D. H. (2007). Testing alternative explanations of phantom decoy effects. Journal of Behavioral Decision Making, 20(3), 323-341.

Default Effect

Definition: The Default Effect is a cognitive bias that describes our tendency to accept the pre-set, default options because, let's be honest, making decisions can be just as exhausting as a high-intensity interval training session.

Description: This bias plays off our natural human tendency to go with the flow - or in this case, the pre-checked box. It's because we have a fondness for the path of least resistance, coupled with a dash of 'if it ain't broke, don't fix it' mentality. It's a potent combo that often leads us to stick with the status quo, unless we're motivated enough to change.

Examples: Politically, this can happen when incumbent politicians are re-elected simply because they're the default choice. In terms of religion, people might stick to their family's faith by default, because questioning or exploring alternatives requires effort. And in social issues, norms and customs often persist because they are the default and changing them requires work.

Impact: The Default Effect can stifle innovation, impede social progress, and create a myopic view of the world. Sometimes the default isn't always the best choice. Imagine if we still defaulted to using rotary phones. Not exactly cutting-edge communication, right?

Related Biases: Status quo bias, system justification, choice-supportive bias.

Countermeasures: Make a habit of questioning the default. Encourage diversity of thought. Seek alternatives.

Imagine if the default was always the opposite—how would that change your decision?

Key Takeaways: Just because it's the default doesn't mean it's the best. It might just be the lazy option. So, go against the current, challenge the norm, and don't let the default effect dull your sparkle.

References:

Johnson, E. J., & Goldstein, D. (2003). Do defaults save lives?. Science, 302(5649), 1338-1339.

Dinner, I., Johnson, E. J., Goldstein, D. G., & Liu, K. (2011). Partitioning default effects: Why people choose not to choose. Journal of Experimental Psychology: Applied, 17(4), 332.

Denomination Effect

Definition: The Denomination Effect is a cognitive bias where people are less likely to spend larger bills compared to smaller ones or coins because, you know, a hundred-dollar bill feels much more valuable than a heap of quarters.

Description: Despite the fact that the face value of money remains the same, regardless of its form, we often attribute a higher psychological value to larger denominations. It's like thinking a chocolate cake is more valuable than the same cake cut into slices. Still the same cake, folks.

Examples: Politically, this could manifest as public opposition to lump sum taxes versus smaller, less noticeable levies. In religious contexts, it might be easier to convince people to contribute smaller, recurring donations than a single large contribution. And in social issues, it may be more palatable to introduce changes incrementally rather than in one fell swoop.

Impact: The Denomination Effect can hinder rational financial decision-making and contribute to inefficiency in spending. Plus, it can subtly influence policy-making, philanthropy, and social change.

Related Biases: Unit bias, anchoring, money illusion.

Countermeasures: Recognize that value is consistent, regardless of denomination or form. And remember, a heap of quarters can buy just as many chocolate cakes as a hundred-dollar bill.

Key Takeaways: Money is money, no matter its shape or size. Don't let the denomination deter you from making rational decisions. After all, you wouldn't eat less cake just because it's cut into slices, right?

References:

Raghubir, P., & Srivastava, J. (2009). The denomination effect. Journal of Consumer Research, 36(4), 701-713.

Disconfirmation Bias

Definition: Disconfirmation bias is the cognitive bias where we tend to scrutinize information that contradicts our beliefs more critically than information that confirms them. It's like being an overly strict judge on a talent show for any idea that sings a different tune to yours.

Description: We are all the champions of our beliefs and when they are challenged, the disconfirmation bias kicks in. We become merciless critics, scrutinizing and discrediting the contradictory information with the energy of a caffeinated detective on a murder mystery.

Examples: In politics, this bias can lead to polarized debates where people are more intent on debunking the opposition's points than considering their merits. In religion, disconfirmation bias can contribute to intolerance and dogmatism, where contradicting beliefs are dismissed without fair consideration. On social issues, this bias can stall progressive dialogues and maintain status quo.

Impact: Disconfirmation bias hinders open-mindedness, inhibits effective communication, and can contribute to social, political, and religious division. In essence, it's a serious roadblock on the highway to mutual understanding.

Related Biases: Confirmation bias, belief bias, biased assimilation.

Countermeasures: Encourage critical thinking and empathy. Strive to give contrasting views a fair trial in the courtroom of your mind before passing the verdict. It might be tough, but hey, nobody said being a fair judge was easy.

Key Takeaways: Don't let disconfirmation bias turn you into a ruthless critic of opposing views. Remember, understanding isn't about agreeing, but about fostering dialogue and broadening perspectives. Keep an open mind, even when the ideas on the stage are singing a different tune.

References:

Taber, C. S., & Lodge, M. (2006). Motivated skepticism in the evaluation of political beliefs. American Journal of Political Science, 50(3), 755-769.

Hart, W., Albarracín, D., Eagly, A. H., Brechan, I., Lindberg, M. J., & Merrill, L. (2009). Feeling validated versus being correct: A meta-analysis of selective exposure to information. Psychological Bulletin, 135(4), 555-588.

Disjunction Fallacy

Definition: The disjunction fallacy is a cognitive bias that occurs when people incorrectly judge that specific conditions are more probable than a more general version of those same conditions. It's like betting on your favorite football player to score a touchdown and win the game, over just your team winning.

Description: This fallacy is a fascinating quirk of our reasoning machinery where specifics seem to take the front seat while the general scenario is left hitchhiking. It's not that we don't like the general scenario; we just find the specific one more enticing, like preferring a chocolate chip cookie over just a cookie.

Examples: In politics, someone might believe it more likely for a candidate to win the election and enact specific policies rather than just winning the election. In religion, a person might feel it's more probable for their deity to respond to a very specific prayer rather than a more general one. In social issues, an activist might believe it's more likely for a specific law to pass and solve a problem, rather than the law just passing.

Impact: The disjunction fallacy can distort our expectations, resulting in flawed decision-making, inaccurate predictions, and frustration when reality doesn't match our very specific expectations.

Related Biases: Conjunction fallacy, availability heuristic.

Countermeasures: Practice critical thinking and remember that more general conditions are always more likely than specific ones, even if the specifics have more details that make them seem vivid or desirable.

Key Takeaways: Don't let the disjunction fallacy lead you down the garden path of unrealistic expectations. Remember that general scenarios are always more likely than their specific counterparts. The cookie, whether it has chocolate chips or not, is always more probable.

References:

Tversky, A., & Kahneman, D. (1983). Extensional versus intuitive reasoning: The conjunction fallacy in probability judgment. Psychological Review, 90(4), 293–315.

Bar-Hillel, M., & Neter, E. (1993). How alike is it versus how likely is it: A disjunction fallacy in probability judgments. Journal of Personality and Social Psychology, 65(6), 1119–1131.

Disregard of Regression Toward the Mean

Definition: The disregard of regression toward the mean is a cognitive bias that occurs when people fail to account for the statistical phenomenon that extreme data tends to get less extreme over time. It's like believing every sequel to a blockbuster movie will be just as great, even when "Return of the Jedi" gently whispers in your ear that this is not always the case.

Description: This bias leans on the failure to recognize that performance, in any field, often follows a bell curve with most results bunching in the middle and fewer at the extremes. However, people usually have their eyes fixed on the extremes and expect them to persist, akin to staring at a solar eclipse and expecting to retain perfect vision.

Examples: In politics, this bias might lead someone to expect a particularly polarizing candidate to maintain their extreme stances. In religion, believers might expect a series of miraculous events to continue indefinitely. On social issues, one might expect unusually high or low crime rates in a given year to persist into the future.

Impact: Disregard of regression toward the mean can lead to unrealistic expectations, inappropriate actions, and disappointment when the "Return of the Jedi" of your expectations turns out to be more of a "Phantom Menace".

Related Biases: Outcome bias, base rate fallacy.

Countermeasures: To mitigate this bias, consider the broader context and remember that extremes are usually followed by more average outcomes, like how a storm is often followed by a calm.

Key Takeaways: Stay realistic and acknowledge the likelihood of regression toward the mean. Remember, not every movie can be "The Empire Strikes Back". Sometimes, you get "The Phantom Menace".

References:

Kahneman, D., & Tversky, A. (1973). On the psychology of prediction. Psychological Review, 80(4), 237–251.

Kahneman, D. (2011). Thinking, fast and slow. Macmillan.

Divergence Bias

Definition: Divergence bias is a cognitive hiccup where individuals overestimate the difference between two options when presented separately compared to when they are presented together. It's like thinking the difference between a banana and an apple is akin to the difference between a bicycle and an airplane, until you see them next to each other in a fruit bowl and suddenly remember what comparisons really mean.

Description: This bias often sneaks up on us in the realm of decision-making, especially when evaluations are carried out separately. When items are evaluated separately, each item is considered in isolation, and we are prone to exaggerate differences. But when we evaluate the same items jointly, side by side, we suddenly become champions of nuance and accuracy.

Examples: In the context of politics, this might manifest when comparing policies of two candidates separately, as opposed to side by side. In religion, interpretations of religious texts might seem drastically different until compared directly. On social issues, two policies may seem miles apart until you see them side by side and realize they both essentially say, "Be kind to each other."

Impact: This bias can lead to decision paralysis, misinformed choices, and distorted perceptions. It can also lead to overblown internet debates where two sides are arguing fiercely over what, upon closer inspection, are relatively minor differences.

Related Biases: Contrast effect, distinction bias.

Countermeasures: To mitigate this bias, consider comparing options side-by-side, rather than separately, and always remember to take a step back and evaluate the real extent of differences. It's not always apples and bicycles; sometimes it's just apples and slightly different apples.

Key Takeaways: Divergence bias can make a molehill seem like a mountain. Keep it in check by comparing things directly and remembering that while all fruit is not created equal, they are all, at the end of the day, still fruit.

References:

Hsee, C. K., & Zhang, J. (2010). General evaluability theory. Perspectives on Psychological Science, 5(4), 343-355.

Hsee, C. K., & Leclerc, F. (1998). Will products look more attractive when presented separately or together? Journal of Consumer Research, 25(2), 175-186.

Door-In-The-Face Technique

Definition: The Door-In-The-Face Technique is a persuasion strategy, not precisely a cognitive bias, that involves making a large request that is likely to be rejected, followed by a smaller, more reasonable request, which seems much nicer in comparison. It's the psychological equivalent of asking someone to build you a spaceship, then following up with "Or could you just pass me the remote?"

Description: The technique operates on the principles of reciprocity and contrast. After denying a large request, individuals often feel obliged to comply with a smaller one. It's like being asked to swim across the Atlantic, saying no, and then feeling pretty okay about swimming across a small pond.

Examples: In politics, a party might propose a radical reform, expecting resistance, only to later propose a more moderate version which is accepted. In religious contexts, a community might initially ask its members for a significant donation and then settle for a smaller one, making the smaller request seem reasonable in comparison. In social issues, advocates might first suggest a drastic measure and then follow it with a less intense one that appears more acceptable.

Impact: This technique can skew decision-making and judgment, making smaller requests or concessions seem more reasonable or acceptable after an initial, larger request has been declined. It's like being relieved when a bear decides to swipe at your picnic basket instead of you – it's still not ideal, but it could have been worse.

Related Biases: Contrast effect, reciprocity norm, anchoring.

Countermeasures: Awareness is the first step. Once you recognize this technique, you can better assess each request on its own merits rather than in comparison to an inflated initial ask. Don't let that first bear swipe ruffle your picnic blanket, so to speak.

Key Takeaways: The Door-In-The-Face technique leverages the power of comparison and reciprocity to make subsequent requests appear more reasonable. To counter it, assess each request independently and don't let that initial spaceship-building proposition sway you.

References:

Cialdini, R. B., Vincent, J. E., Lewis, S. K., Catalan, J., Wheeler, D., & Darby, B. L. (1975). Reciprocal concessions procedure for inducing compliance: The door-in-the-face technique. Journal of Personality and Social Psychology, 31(2), 206.

Fernandez, S., & Bevacqua, J. (1982). Extending the door-in-the-face compliance strategy to a commercial request. Journal of Consumer Research, 9(1), 96-100.

Dunning-Kruger Effect

Definition: The Dunning-Kruger Effect is a cognitive bias wherein people with low ability at a task overestimate their ability, and conversely, individuals with high ability underestimate their competence.

Description: Named after psychologists David Dunning and Justin Kruger, who identified the bias in a series of experiments in 1999, the Dunning-Kruger Effect involves a metacognitive inability of unskilled individuals to recognize their ineptitude and evaluate their ability accurately. The result is a dangerously comical paradox where some of the least competent folks fancy themselves experts, while the truly talented underestimate their mastery. Talk about an intellectual version of a cat chasing its tail.

Examples: You might see this at play in a political debate, where an individual with minimal understanding of complex policies passionately believes they have the solutions to major societal problems, while experienced politicians may underestimate their understanding of the same issues. A case of "I read one article on the internet, and now I am an expert."

Impact: The Dunning-Kruger Effect can impede learning and growth, as people fail to recognize their mistakes or gaps in their knowledge. It can also contribute to polarization in social and political debates, as those with superficial understanding often express the most extreme confidence in their views. So, it's not just damaging on a personal level; it can escalate societal disputes like gasoline on a dumpster fire.

Related Biases: The Dunning-Kruger Effect is related to the bias blind spot (the failure to see one's own biases) and the overconfidence effect (excessive certainty in the accuracy of one's beliefs).

Countermeasures: To combat the Dunning-Kruger Effect, fostering an attitude of intellectual humility and openness to feedback can be effective. Encouraging lifelong learning and teaching metacognitive skills are also beneficial. Or in plain English: Know that you don't know everything, listen when others tell you so, and never stop learning.

Key Takeaways: The Dunning-Kruger Effect is a potent reminder of the importance of humility, self-awareness, and continuous learning. Remember, the first step towards being sort of good at something is sucking at it but having absolutely no idea that you do.

References:

Dunning, D., Johnson, K., Ehrlinger, J., & Kruger, J. (2003). Why people fail to recognize their own incompetence. Current Directions in Psychological Science, 12(3), 83-87.

Kruger, J., & Dunning, D. (1999). Unskilled and unaware of it: how difficulties in recognizing one's own incompetence lead to inflated self-assessments. Journal of Personality and Social Psychology, 77(6), 1121.

Duration Neglect

Definition: Duration Neglect is a cognitive bias in which the duration of an event is ignored when evaluating and remembering the event's overall experience. The intensity of the peak (most intense) and the end of the experience tend to dominate the memory of the event, regardless of its duration.

Description: Coined by Nobel laureate Daniel Kahneman and his collaborator Amos Tversky, duration neglect is the estranged cousin of time-related cognitive biases. With this bias, whether your tedious board meeting lasted for an hour or the whole afternoon, you remember it with equal disdain. In essence, our brains aren't great at remembering the "quantity" of experiences, whether good or bad, and are more influenced by the "quality" or intensity at its peak and end.

Examples: In a political context, consider a lengthy presidential term characterized by middling success but a disastrous final year due to a crisis. Voters, influenced by duration neglect, may overwhelmingly remember the term as unsuccessful, disregarding the many previous years of competent leadership. It's like forgetting all the good parts of a vacation just because you lost your luggage on the last day.

Impact: Duration Neglect can lead to misjudgments in personal decisions and skewed perceptions of events. It can influence how we perceive social issues, political terms, or even our own life experiences. It can also unfairly skew public opinion, tainting reputations or influencing election outcomes. Simply put, it can make mountains out of molehills and reduce mountains to molehills.

Related Biases: Duration Neglect is a component of the broader "Peak-End Rule" bias, which asserts that people judge an experience largely based on how they felt at its peak and at its end. It is also related to the Recency Effect, where most recent information is remembered best.

Countermeasures: Becoming aware of the bias and practicing mindfulness can help us appreciate the full scope of an experience, not just its peak or end. For decision-making, it can be beneficial to use systematic and structured approaches that consider the whole picture and not just salient moments.

Key Takeaways: Duration Neglect is a stark reminder that, in memory, all moments are not created equal. Being aware of this can help you in forming more accurate judgments and developing a fuller appreciation for the entirety of experiences.

References:

Kahneman, D., Fredrickson, B. L., Schreiber, C. A., & Redelmeier, D. A. (1993). When More Pain Is Preferred to Less: Adding a Better End. Psychological Science, 4(6), 401–405.

Fredrickson, B. L., & Kahneman, D. (1993). Duration Neglect in Retrospective Evaluations of Affective Episodes. Journal of Personality and Social Psychology, 65(1), 45–55.

Effort Justification

Definition: Effort Justification is a cognitive bias in which individuals attribute greater value to an outcome they worked hard to achieve, even if the outcome may not justify the effort expended. Essentially, it's the brain's way of saying, "I worked hard for this, so it has to be good!"

Description: "No pain, no gain" may be a great slogan for gym rats, but it's also a potent illustration of the Effort Justification bias. It's the mental accounting trick we play on ourselves when we invest a substantial amount of effort, time, or money into something. Rather than face the potentially bitter fact that our effort was not worth the outcome, we pump up the perceived value of the outcome.

Examples: An example from the realm of politics might be a voter who spends hours campaigning for a candidate. Despite the candidate's subsequent lackluster performance in office, the voter continues to believe that they were the best choice. Or, think of the followers of a religion who undergo strenuous rites or rituals and subsequently report a deepened belief in their faith, regardless of the objective benefits or drawbacks of these practices.

Impact: Effort Justification can distort our perception of the value or quality of an outcome, leading us to make irrational judgments and potentially invest even more resources into endeavors that may not be worthwhile. It can perpetuate unproductive behaviors and reinforce questionable beliefs.

Related Biases: This bias is closely tied to Cognitive Dissonance, a state of mental discomfort that arises when our actions and beliefs do not align. Effort Justification can be seen as a specific manifestation of Cognitive Dissonance.

Countermeasures: Recognizing and acknowledging this bias is the first step. Ask yourself if the outcome truly matches the effort or if you are justifying a disproportionate effort. Seek objective feedback or use empirical measures to assess outcomes, when possible.

Key Takeaways: Don't let the sweat, tears, or blood (figuratively, we hope!) fool you into overvaluing the fruits of your labor. Effort is important, but it doesn't inherently increase the worth of an outcome.

References:

Aronson, E., & Mills, J. (1959). The effect of severity of initiation on liking for a group. Journal of Abnormal and Social Psychology, 59, 177–181.

Festinger, L., & Carlsmith, J. M. (1959). Cognitive consequences of forced compliance. The Journal of Abnormal and Social Psychology, 58(2), 203–210.

Egocentric Bias

Definition: The Egocentric Bias is the cognitive tendency to overestimate one's contribution to a group task or to believe that others share the same perspective and beliefs. In a nutshell, it's the universe's sense of humor – making us all believe we're the center of it.

Description: Just as the Earth once basked in the erroneous belief that it was the center of the universe, humans, too, often fall prey to the deceptive allure of egocentric bias. This bias occurs when we overvalue our own perspective, largely ignoring or misunderstanding the views, thoughts, and feelings of others.

Examples: Consider a political activist who assumes everyone on their social media feed agrees with their passionate posts about a particular candidate or policy. Or the religious leader who firmly believes that their personal interpretations of sacred texts are universally shared among the congregation. Even that delightful person who monopolizes the conversation at a social gathering, convinced their stories are the life of the party, is showcasing the egocentric bias.

Impact: The Egocentric Bias can create misunderstandings and conflicts in interpersonal relationships, hinder collaboration, and contribute to a polarized society where everyone is shouting, and no one is listening. It can lead to rash decisions based on the erroneous belief that everyone agrees with us or values the same things we do.

Related Biases: Egocentric Bias shares close ties with the False Consensus Effect (overestimating how much others agree with us) and the Fundamental Attribution Error (blaming people's behavior on their personality rather than situational factors).

Countermeasures: Actively seeking out and considering diverse perspectives can help combat this bias. Regularly practicing empathy and trying to understand the experiences and viewpoints of others can also go a long way.

Key Takeaways: Just remember, even if you're the star of your own life, you're not necessarily the leading role in everyone else's. Keep your Egocentric Bias in check, or you might just find yourself playing the villain instead.

References:

Ross, L., Greene, D., & House, P. (1977). The "false consensus effect": An egocentric bias in social perception and attribution processes. Journal of Experimental Social Psychology, 13(3), 279–301.

Kruger, J., & Gilovich, T. (1999). "Naive cynicism" in everyday theories of responsibility assessment: On biased assumptions of bias. Journal of Personality and Social Psychology, 76(5), 743–753.

Endowment Effect

Definition: The Endowment Effect is the cognitive bias that causes people to overvalue something simply because they own it. It's the universe's sly way of making us treasure the most ridiculous items, just because they are ours.

Description: Coined by Richard Thaler, the endowment effect reveals how possession significantly increases our valuation of an item. It's a bit like parents believing their children are more brilliant than anyone else's, but with objects. Ownership, it appears, fosters emotional attachment, leading us to set unrealistic prices for items we own, regardless of their actual market value.

Examples: Ever been to a garage sale where a tattered teddy bear is priced like a rare artifact? Or how about that political supporter who would never part with their signed campaign poster, even for a hefty sum? That's the Endowment Effect in action.

Impact: The Endowment Effect can lead to irrational financial decisions and economic inefficiencies. It can hinder negotiations or trades and cause us to over-invest in items (or beliefs) simply because they're "ours." It also influences policy and law, such as in cases of eminent domain where people demand compensation far above market value for their properties.

Related Biases: The Endowment Effect is closely related to Loss Aversion (the tendency to prefer avoiding losses to acquiring equivalent gains) and the Status Quo Bias (a preference for the current state of affairs).

Countermeasures: Recognizing the Endowment Effect is the first step towards mitigating it. Try to objectively assess the value of your possessions, or better yet, ask an unbiased party for their perspective.

Key Takeaways: While it's nice to treasure your belongings, don't fall for the seductive whisper of the Endowment Effect convincing you that your old toaster is worth its weight in gold. Remember, ownership can inflate perceived value, but the market may beg to differ.

References:

Thaler, R. (1980). Toward a positive theory of consumer choice. Journal of Economic Behavior & Organization, 1(1), 39–60.

Morewedge, C. K., & Giblin, C. E. (2015). Explanations of the endowment effect: an integrative review. Trends in cognitive sciences, 19(6), 339-348.

Epistemic Closure

Definition: Epistemic Closure is the philosophical concept where a belief system is 'closed' to counter-evidence, making its adherents less likely to engage with information contradicting their views. It's like being stuck in an echo chamber where the only voice you hear is your own, or worse, one that keeps reaffirming your own views.

Description: This concept involves creating a knowledge system that's airtight to any information that challenges its existing beliefs. It's a psychological security blanket, keeping out the cold drafts of contrasting views. In a nutshell, you know what you know, and you're not interested in knowing anything else.

Examples: If you've ever tried arguing with someone who firmly believes that their political party can do no wrong, or with a religious zealot who dismisses all evidence contradicting their faith, you've danced with Epistemic Closure.

Impact: Epistemic Closure can lead to polarization in society, especially in areas like politics and religion, where people become more entrenched in their beliefs and less likely to engage in productive dialogue. It also stifles intellectual growth, hindering problem-solving, and the development of well-rounded viewpoints.

Related Biases: Confirmation Bias (favoring information that confirms one's existing beliefs), Belief Perseverance (maintaining beliefs despite contradicting evidence), and Echo Chamber Effect (where shared ideas are amplified and reinforced in a closed system).

Countermeasures: Encourage critical thinking and exposure to diverse viewpoints. Make a point of seeking out and considering contradictory evidence. Foster an open, curious mindset, and remember that certainty can sometimes be the enemy of wisdom.

Key Takeaways: Epistemic Closure might feel comforting, but it's like living in a self-constructed bubble – it's limiting and brittle. Open the doors, let in some fresh ideas, and remember – even the most comfortable echo chamber is still just a chamber.

References:

Kahan, D. M. (2015). Climate-science communication and the measurement problem. Advances in Political Psychology, 36, 1-43.

Nickerson, R. S. (1998). Confirmation bias: A ubiquitous phenomenon in many guises. Review of General Psychology, 2(2), 175-220.

Experimenter's Bias (Expectation Bias)

Definition: Expectation Bias is a cognitive quirk where people's expectations influence their perception of reality. It's the mental equivalent of wearing rose-tinted glasses that match the color of your presumptions.

Description: This bias occurs when we unconsciously influence our perceptions and interpretations to fit our expectations. It's like having a hunch, then bending over backwards to make sure that hunch comes true. We don't just passively perceive the world; our brains are active interpreters, always trying to make the facts fit the narrative.

Examples: Ever expected a movie to be dreadful, only to find it surprisingly tolerable? Or maybe you've believed a politician's promise, despite a track record suggesting you shouldn't. In both scenarios, Expectation Bias is at work, subtly shaping your perception and interpretation.

Impact: In politics, religion, and social issues, Expectation Bias can lead to false assumptions and flawed decisions. It can cause us to ignore important evidence, dismiss valuable viewpoints, and maintain harmful stereotypes.

Related Biases: Confirmation Bias (favoring information that aligns with our existing beliefs), Observer-Expectancy Effect (influencing a study's results in line with the researcher's expectations), and Optimism Bias (overestimating the probability of positive outcomes).

Countermeasures: Foster a mindset of critical thinking and open-mindedness. Make an effort to test your assumptions and seek out evidence that contradicts your expectations. Also, practicing mindfulness can help increase awareness of how your expectations might be coloring your perceptions.

Key Takeaways: Expectations are powerful and can subtly shape our perceptions. Recognizing this can help you stay open to new information and make better, more informed decisions. Remember, your mental tinted glasses might make the world look rosier, but they can also blind you to reality.

References:

Pannucci, C. J., & Wilkins, E. G. (2010). Identifying and avoiding bias in research. Plastic and Reconstructive Surgery, 126(2), 619-625.

Stangor, C., & McMillan, D. (1992). Memory for expectancy-congruent and expectancy-incongruent information: A review of the social and social developmental literatures. Psychological Bulletin, 111(1), 42-61.

Extension Neglect

Definition: Extension Neglect is a cognitive hiccup that causes individuals to underestimate the size or scope of a set or category, particularly when it involves abstract rather than concrete elements. It's like trying to count the stars and giving up after a hundred, concluding, "That's probably about it."

Description: This bias originates from our difficulty in comprehending large quantities or conceptual entities. We underestimate the number of people affected by a policy, the number of adherents of a religion, or the diversity within a political party. Basically, we tend to forget that a spoonful of data doesn't make an ocean of understanding.

Examples: Ever thought all supporters of a political party share the exact same views? Or that all members of a religion follow the same practices? These over-generalizations are examples of Extension Neglect in action, grossly underestimating the diversity within these groups.

Impact: Extension Neglect can lead to oversimplification of complex issues and propagation of stereotypes in religion, politics, and social discourse. It reduces our ability to engage in nuanced thinking and hampers effective decision-making.

Related Biases: Base Rate Fallacy (disregarding statistical rates in favor of specific information), Stereotyping (assuming all members of a group share the same characteristics), and The Law of Small Numbers (overemphasizing small samples).

Countermeasures: Cultivating awareness of this bias is the first step. Practice thinking in terms of probabilities and ranges rather than fixed quantities or characteristics. Make an effort to explore the complexity and diversity within large groups or sets.

Key Takeaways: Life, especially when it comes to understanding complex groups and concepts, is not a 'one-size-fits-all' proposition. Remember, while it's easier to fit the universe into a teacup, you're going to miss out on a whole lot of stars.

References:

Tversky, A., & Kahneman, D. (1974). Judgment under uncertainty: Heuristics and biases. Science, 185(4157), 1124-1131.

Tversky, A., & Koehler, D. J. (1994). Support theory: A nonextensional representation of subjective probability. Psychological Review, 101(4), 547-567.

Extrinsic Incentives Bias

Definition: Extrinsic Incentives Bias is our tendency to overestimate the influence of external rewards and underestimate the role of intrinsic motivation in people's actions. It's the cousin of the 'Money-is-the-root-of-all-actions' philosophy.

Description: This bias stems from our focus on tangible rewards as motivators. We overlook the fact that people often act out of passion, interest, or moral conviction. It's like thinking your friend attends Sunday church solely for the free cookies, ignoring their profound spiritual engagement.

Examples: A politician passionately advocating for climate change policies might be seen as vying for popularity, disregarding their potential intrinsic concern for the environment. Similarly, an individual advocating for social justice might be misinterpreted as seeking social media fame, overlooking their possible deep-rooted commitment to equality.

Impact: This bias can lead to misunderstandings, misconceptions, and faulty evaluations in religious, political, and social contexts. It can hamper trust and cooperation, and can often result in suboptimal strategies for motivation.

Related Biases: Fundamental Attribution Error (overemphasis on personal characteristics and underestimation of situational factors) and Over-justification Effect (external rewards diminishing intrinsic motivation).

Countermeasures: Be aware of this bias and consider the possibility of intrinsic motivation. Avoid quick judgments and encourage open dialogue to understand others' motivations better.

Key Takeaways: Actions often stem from a complex interplay of intrinsic and extrinsic motivations. Remember, people are not vending machines that operate solely on the insertion of tangible rewards.

References:

Frey, B. S., & Jegen, R. (2001). Motivation crowding theory. Journal of Economic Surveys, 15(5), 589-611.

Deci, E. L., Koestner, R., & Ryan, R. M. (1999). A meta-analytic review of experiments examining the effects of extrinsic rewards on intrinsic motivation. Psychological Bulletin, 125(6), 627-668.

False Consensus Effect

Definition: The False Consensus Effect refers to our proclivity to overestimate the degree to which others agree with us. It's a bit like assuming everyone must adore pineapple on pizza because, well, *you* do.

Description: This bias is deeply rooted in our desire to conform and feel "normal". We tend to believe our views, attitudes, and behavior are typical, causing us to overestimate their prevalence in others. It's like believing everyone secretly enjoys watching cat videos in their spare time just because you do.

Examples: A religious individual might assume that their religious beliefs are more widely shared than they actually are. Politically, someone might overestimate the number of people who share their political leanings, leading to surprise when election results don't align with their expectations. Socially, one might think that their opinion about a controversial topic, like pineapple on pizza, is the popular view.

Impact: The False Consensus Effect can distort our understanding of diverse perspectives in religion, politics, and social issues. It can lead to polarization, misunderstandings, and can hinder constructive dialogue.

Related Biases: Confirmation Bias (favoring information that confirms our pre-existing beliefs) and Egocentric Bias (overemphasis on our own perspective and experiences).

Countermeasures: Foster a mindset of curiosity and openness, actively seeking diverse perspectives. Conduct reliable surveys to gather data instead of relying on personal hunches.

Key Takeaways: Just because you enjoy pineapple on pizza doesn't mean everyone does. Embrace diverse perspectives and understand that consensus is not always as widespread as it appears to be.

References:

Ross, L., Greene, D., & House, P. (1977). The "false consensus effect": An egocentric bias in social perception and attribution processes. Journal of Experimental Social Psychology, 13(3), 279-301.

Mullen, B., Atkins, J. L., Champion, D. S., Edwards, C., Hardy, D., Story, J. E., & Vanderklok, M. (1985). The false consensus effect: A meta-analysis of 115 hypothesis tests. Journal of Experimental Social Psychology, 21(3), 262-283.

False Dichotomy

Definition: False Dichotomy, sometimes known as a False Dilemma or either/or fallacy, is the cognitive bias where only two options are considered, when in fact there are more. It's like thinking the only pizza toppings are pineapple or ham, ignoring the vast universe of potential toppings.

Description: This bias often occurs when we oversimplify a complex issue into a binary choice. It's a cognitive shortcut we take when the world's complexity is a bit too daunting, or when we want to steer people toward a particular choice.

Examples: Politically, we often reduce complex matters into binary choices: you're either conservative or liberal, capitalism or socialism. Religiously, you're either a believer or an atheist. Socially, you're either pro-pineapple on pizza or against it, completely neglecting the mushroom and olive lovers among us.

Impact: False Dichotomy can polarize debates, oversimplify complex issues, and close the door to potential solutions or compromises. It reduces the richness of the world's diversity into a one-or-the-other perspective.

Related Biases: Black-and-White Thinking (viewing things in absolute terms) and Overconfidence Bias (overestimating the accuracy of our beliefs).

Countermeasures: Recognize the complexity of most issues and actively explore more than two options. Promote nuanced thinking and understand that most things exist on a spectrum rather than in binary form.

Key Takeaways: The world is not just black and white, and pizza toppings are not just ham or pineapple. Embrace the full spectrum of possibilities and resist the urge to oversimplify.

References:

Baron, J. (2008). Thinking and deciding (4th ed.). Cambridge University Press.

Thompson, V. A. (2009). Dual process theories: A metacognitive perspective. In J. Evans & K. Frankish (Eds.), In two minds: Dual processes and beyond (p. 171–195). Oxford University Press.

False Fame Effect

Definition: The False Fame Effect is the cognitive bias where we mistake familiarity for fame. That is, you think you've heard of "John Doe" because he's famous when, in fact, he's just an average guy whose name you've heard recently.

Description: This bias shows up when we're exposed to a name or a face and later mistake that familiarity for some sort of significance. It's like seeing the mailman at the supermarket and, for a fleeting second, thinking you've spotted a celebrity—only to realize that he's the guy who delivers your Amazon packages.

Examples: In the realm of social issues, this can be seen with activism, where the most familiar names are often seen as the most influential, regardless of their actual impact. In politics, this bias might explain why incumbents have an advantage in elections, even when their performance has been less than stellar.

Impact: The False Fame Effect can lead us to overvalue familiar but irrelevant information, potentially influencing our decisions and judgments. If we're not careful, we could end up electing a ham sandwich as president simply because we've heard its name a few times.

Related Biases: Availability Heuristic (relying on immediate examples when evaluating a specific topic or decision), Illusory Truth Effect (believing information to be correct after repeated exposure).

Countermeasures: It's important to critically evaluate why we feel someone or something is significant. Is it based on real accomplishments and merit, or just familiarity?

Key Takeaways: Familiarity doesn't equal fame or significance. So, before you ask your mailman for an autograph, make sure he's actually someone famous.

References:

Jacoby, L. L., Kelley, C., Brown, J., & Jasechko, J. (1989). Becoming famous overnight: Limits on the ability to avoid unconscious influences of the past. Journal of Personality and Social Psychology, 56(3), 326-338.

Whittlesea, B. W., Jacoby, L. L., & Girard, K. (1990). Illusions of immediate memory: Evidence of an attributional basis for feelings of familiarity and perceptual quality. Journal of Memory and Language, 29(6), 716-732.

False Memory

Definition: False Memory is a cognitive bias where we recall events or details that never actually happened. Yes, that's right, our brain is quite the prolific fiction writer, often spicing up our personal histories with tantalizing bits of non-reality.

Description: This bias revolves around the fragility and malleability of our memories. Our brain is not a perfect recorder. It is more like an overenthusiastic painter, adding a dash of color here, a shadow there, and sometimes a whole new character to the canvas of our past.

Examples: In politics, this can lead to inaccurate recollections of past events or policies, creating a skewed perspective of reality. In religion, false memories might contribute to the creation and reinforcement of certain beliefs or rituals. For social issues, it can distort perceptions and fuel controversies.

Impact: False Memory can lead to misguided decisions based on inaccurate information. It can distort historical understanding, skew perspectives, and lead to unwarranted actions. So, before you swear that your chosen political candidate was a war hero because you "remember" him being so, it might be worth double-checking your sources.

Related Biases: Confirmation Bias (favoring information that confirms our pre-existing beliefs), Misinformation Effect (memories being influenced by incorrect or misleading information post-event).

Countermeasures: Critical thinking, mindfulness, and fact-checking can help. Before making important decisions based on memory, verify the accuracy of the recollection.

Key Takeaways: Our memories are more akin to a game of 'Telephone' than a perfect recording. Be cautious about the accuracy of recalled events, particularly when these memories are driving your decisions.

References:

Loftus, E. F. (1997). Creating false memories. Scientific American, 277(3), 70-75.

Roediger, H. L., & McDermott, K. B. (1995). Creating false memories: Remembering words not presented in lists. Journal of Experimental Psychology: Learning, Memory, and Cognition, 21(4), 803.

Fading Affect Bias

Definition: Fading Affect Bias is the phenomenon where negative emotions associated with past events fade faster than positive ones. In simple words, our brain is like a soft-hearted grandparent, choosing to remember our childhood mischief fondly while conveniently forgetting the tantrums.

Description: This bias is the brain's attempt to maintain emotional balance. It might be nature's way of ensuring we don't spend our lives wallowing in the puddles of past miseries. After all, who wants to remember every detail of the time they accidentally called their third-grade teacher "mom"?

Examples: In the realm of politics, voters might remember a leader's accomplishments more vividly than their scandals over time. In religious contexts, the positive aspects of rituals and communal gatherings might stick around longer than the less pleasant ones. Socially, it can lead to a rosier view of past events or societal norms.

Impact: The Fading Affect Bias can alter our perceptions of the past, influencing future decisions and judgments. It can impact how we vote, which practices we follow, and how we perceive societal norms. It's why your teenage rebellion phase might look like a charming rite of passage in retrospect.

Related Biases: Positivity Bias (the tendency to focus on positive aspects of life), Rosy Retrospection (remembering past times more positively than they were).

Countermeasures: Awareness is the first step. Recognize that your recollections may be sugar-coated. Balanced reflection and external information can help provide a more accurate perspective.

Key Takeaways: Our emotional memories are akin to old photographs; they fade over time, and often, the unpleasant bits fade faster. When reflecting on the past, be aware that your memory might be playing you a sweet, deceitful song.

References:

Ritchie, T. D., Batteson, T. J., Bohn, A., Crawford, M. T., Ferguson, G. V., Schrauf, R. W., Vogl, R. J., & Walker, W. R. (2015). A pancultural perspective on the fading affect bias in autobiographical memory. Memory, 23(2), 278-290.

Walker, W. R., Skowronski, J. J., & Thompson, C. P. (2003). Life is pleasant—and memory helps to keep it that way! Review of General Psychology, 7(2), 203.

Fear of Missing Out (FOMO)

Definition: Fear of Missing Out, fondly known as FOMO, is the dread that others might be having rewarding experiences from which one is absent. It's the psychological equivalent of staring at the last piece of cake while someone else makes a move for it.

Description: Rooted in the social need to belong, FOMO nudges us to say "yes" to everything, attend every party, and scroll social media until our thumbs cramp, lest we miss something supposedly life-changing. It's like a cognitive party-pooper, reminding us there could be a better party happening somewhere else.

Examples: Politics sees a lot of FOMO when supporters fear missing out on voting for the "winning" candidate. Religion can invoke FOMO in those concerned about missing out on the supposed benefits of certain practices. Social media, however, is the FOMO poster child, enticing us with snapshots of others' 'perfect' lives.

Impact: FOMO can lead to increased stress, decision fatigue, and even reckless actions in an attempt to be everywhere and do everything. It's why you may find yourself at a silent disco yoga class at 7 AM on a Sunday, questioning your life choices.

Related Biases: Availability Cascade (a belief becomes more accepted as it is seen more), Bandwagon Effect (people do things because others are doing them).

Countermeasures: Practice mindfulness, focus on quality over quantity of experiences, and take social media with a grain of salt. Remember, most people only post their highlights reel, not their blooper tapes.

Key Takeaways: FOMO is the nagging whisper that the grass is greener on the other side. Cultivate your own garden instead, and remember that sometimes the most rewarding experience is a good night's sleep.

References:

Przybylski, A. K., Murayama, K., DeHaan, C. R., & Gladwell, V. (2013). Motivational, emotional, and behavioral correlates of fear of missing out. Computers in Human Behavior, 29(4), 1841-1848.

Wortham, J. (2011, April 10). Feel like a wall flower? Maybe it's your Facebook wall. The New York Times, p. 1.

Focusing Effect

Definition: The Focusing Effect is the cognitive bias that has us amplifying the importance of one single factor while undervaluing other relevant factors in decision-making. It's a bit like focusing on a single tree while ignoring the rest of the forest (which could be on fire, by the way).

Description: This bias influences us to overestimate the effect of one characteristic or piece of information when making decisions or predicting outcomes. In essence, it's our brain's "can't see the forest for the trees" moment, and trust me, it's not as poetic as it sounds.

Examples: In politics, voters may focus too much on a candidate's charisma while ignoring their policies. In religion, an individual might concentrate on a single virtue or belief to the detriment of the bigger picture. Socially, it's like deciding to marry someone based solely on their Netflix binge-watching compatibility.

Impact: The Focusing Effect can lead to poor decision-making and skewed perceptions, which could potentially result in regret or disappointment. No one wants to find out their Netflix soulmate only watches horror films at 2 AM.

Related Biases: Anchoring Bias (relying too heavily on initial information), Confirmation Bias (favoring information that confirms existing beliefs).

Countermeasures: Broaden your perspective. Take into account various factors before making a decision. Get a second opinion if needed. And maybe ask your potential spouse about their viewing habits before the wedding.

Key Takeaways: The Focusing Effect is like tunnel vision for decision-making. Widen your field of view and consider all aspects before taking the leap, whether it's voting, religious practices, or Netflix-based marriages.

References:

Schkade, D. A., & Kahneman, D. (1998). Does living in California make people happy? A focusing illusion in judgments of life satisfaction. Psychological Science, 9(5), 340-346.

Kim, H., & John, D. R. (2008). Consumer response to brand extensions: Construal level as a moderator of the importance of perceived fit. Journal of Consumer Psychology, 18(2), 116-126.

Foot-In-The-Door Technique

Definition: The Foot-In-The-Door Technique is a persuasion tactic involving getting someone to agree to a small initial request, thus increasing the likelihood of them agreeing to a larger request later. It's like getting someone to hold the door open for you, then convincing them to carry your groceries, water your plants, and maybe renovate your kitchen.

Description: This technique operates on the principle of consistency; once someone has agreed to a small request, they're more likely to comply with a larger one to stay consistent with their initial agreement. The human brain, being the fan of shortcuts it is, uses past decisions to inform future ones, which can lead to some...interesting results.

Examples: Politically, it could be a party asking you for a small donation, only to request a bigger one later. In religion, you might be asked to attend a small group meeting, which then turns into regular service attendance. And socially, it might be a friend asking to borrow a book, then ending up with your whole library.

Impact: This technique can lead to manipulation and exploitation in situations where there's an imbalance of power or information. So, you might end up being overly generous with your political donations, church commitments, or book collections.

Related Biases: Commitment and Consistency Bias (we strive to be consistent with our past commitments), Reciprocity Bias (we tend to want to return a favor).

Countermeasures: Be aware of the technique and consider each request independently. Take the time to deliberate whether a larger request aligns with your interests and values. And maybe have a lending policy for your books.

Key Takeaways: The Foot-In-The-Door Technique uses our own consistency against us. Always remember to evaluate each request individually. And seriously, watch your book collection.

References:

Freedman, J. L., & Fraser, S. C. (1966). Compliance without pressure: The foot-in-the-door technique. Journal of Personality and Social Psychology, 4(2), 195–202.

Burger, J. M. (1999). The foot-in-the-door compliance procedure: A multiple-process analysis and review. Personality and Social Psychology Review, 3(4), 303–325.

Forer Effect (Barnum Effect)

Definition: The Forer Effect, or Barnum Effect, is the tendency to see vague, general statements as exceptionally accurate descriptions of oneself. It's like going to a fortune teller and believing they know you better than your spouse after a generic "You have a great need for other people to like and admire you" statement.

Description: The underpinning here is our desire to believe in personal validations that boost our self-esteem. We are amazingly good at filling in blanks to suit our needs, even when the 'fill' is as generic as a made-in-China T-shirt.

Examples: Astrology columns that vaguely tell you how your week will be based on your star sign are classic Forer playgrounds. Likewise, the rhetoric in political speeches designed to appeal to "the common man" often smack of the Barnum effect. In social settings, "cold reading" techniques used in psychic readings and mentalism acts often exploit this bias.

Impact: We may make decisions based on false premises, thinking they are tailored to us when they are, in fact, generic. This can lead to misinformed decisions in politics, religion, and even personal relationships.

Related Biases: Confirmation Bias (our tendency to cherry-pick information that confirms our existing beliefs), and the Self-Serving Bias (our knack for attributing positive events to our own character but blaming negative events on external factors).

Countermeasures: Always ask for specificity. Generic statements that can apply to anyone are not necessarily applicable to you. Practicing critical thinking and maintaining a healthy level of skepticism can help, too. And perhaps take your horoscope with a pinch of salt.

Key Takeaways: The Forer Effect illustrates our vulnerability to flattery and the desire for self-validation. Be critical of generic statements that seem to resonate with you personally. Remember, not every "you're unique" statement is unique to you.

References:

Forer, B. R. (1949). The fallacy of personal validation: a classroom demonstration of gullibility. Journal of abnormal psychology, 44(1), 118.

Fichten, C. S., & Sunerton, B. (1983). Popular horoscopes and the "Barnum effect". The Journal of Psychology, 114(1), 123-134.

Framing Effect

Definition: The Framing Effect refers to our brain's little magic trick where it jumps to different conclusions based on how information is presented or "framed."

Description: The Framing Effect arises from the fact that humans, in all our rational glory, still tend to focus more on how something is said rather than what is said. Different words, context, or presentation can influence our perceptions and lead us to different conclusions.

Examples: If a politician says their policy will result in "90% employment," it sounds fantastic, doesn't it? But if they frame it as "10% unemployment," the crowd might start sharpening their pitchforks, despite the fact that the statistics are identical. Similarly, a "95% fat-free" label on a food product sounds healthier than "contains 5% fat", though the reality remains the same.

Impact: The Framing Effect has a significant impact on decision making, often leading to irrational choices in social, political, and religious contexts. Misleading framing can sway public opinion, skew perceived realities, and even influence election results.

Related Biases: The Anchoring Bias (the tendency to rely too heavily on the first piece of information encountered), Confirmation Bias (tendency to interpret information in a way that confirms one's preconceptions), and the Availability Heuristic (making decisions based on readily available information).

Countermeasures: To counter this bias, it's important to consider the underlying facts, not just their presentation.

Always question the information you're given: Is there another way to frame this? What is not being said? Developing strong critical thinking skills can help keep the Framing Effect in check.

Key Takeaways: The Framing Effect shows us that presentation matters. Be aware of how information is framed and ensure you're making decisions based on facts, not just their packaging. As they say, don't judge a book by its cover—or in this case, a fact by its frame.

References:

Tversky, A., & Kahneman, D. (1981). The framing of decisions and the psychology of choice. Science, 211(4481), 453-458.

Levin, I. P., Schneider, S. L., & Gaeth, G. J. (1998). All frames are not created equal: A typology and critical analysis of framing effects. Organizational Behavior and Human Decision Processes, 76(2), 149-188.

Frequency Illusion

Definition: The Frequency Illusion, also known as the Baader-Meinhof phenomenon, is a cognitive bias where a concept or thing you just found out about suddenly seems to appear everywhere.

Description: This charming quirk of the human mind occurs in two steps: First, there's selective attention, where your noggin perks up at something new or interesting; then, confirmation bias takes the wheel, asserting that yes, this thing is now everywhere. In reality, the frequency of these instances hasn't changed—you're just noticing it more.

Examples: Suppose you hear a relatively unknown political term or religious doctrine for the first time. Suddenly, you find it popping up in newspaper articles, social media feeds, and late-night comedy shows. Or maybe you discover a vintage car model, only to see it three times on your way home. Don't worry, you're not being stalked by a car—it's just the Frequency Illusion at work.

Impact: This bias can make us overestimate the prevalence of certain ideas or items, thereby influencing our decisions and judgments. Politically, it may lead to the false consensus effect, making us think our opinions are more common than they actually are.

Related Biases: Confirmation bias, Selective Perception, and Observer-expectancy Effect are close relatives of the Frequency Illusion—they all reflect our brain's uncanny knack for seeing what it wants to see.

Countermeasures: Awareness is the first step. Recognizing the Frequency Illusion when it happens can help keep it in check. Also, when you think something is suddenly "everywhere," remind yourself that it's probably your heightened awareness, not an actual increase in frequency.

Key Takeaways: The Frequency Illusion shows us that our perception isn't always reality. Remember, just because you're suddenly seeing something everywhere doesn't mean it's taking over the world—your brain just bought a new pair of rose-colored glasses.

References:

Zwicky, A. (2006). Just Between Dr. Language and I. In Arnold Zwicky, (Ed.), Two blogs: A diary and The search for the perfect language. Lulu Press.

Mullen, B., Bryant, B., & Driscoll, D. (1997). Cumulative meta-analysis of the relationship between useful intergroup contact and outgroup stereotyping. Basic and Applied Social Psychology, 19(1), 3-16.

Functional Fixedness

Definition: Functional Fixedness is a cognitive bias that hinders one's ability to use an object in a novel way. It keeps us locked into the idea that an object can only be used in the way it was traditionally intended.

Description: Picture your mind as a super efficient filing system. When it encounters an object—say, a paperclip—it immediately files it under "Things Used to Clip Papers Together." Now, when you need a makeshift lock pick or a tiny coil, your mind is already entrenched in the idea of the paperclip's paper-clipping destiny. This is Functional Fixedness.

Examples: Take the example of a political rally. The microphone fails, and there's a spatula in the kitchen. Functional Fixedness would prevent you from considering using the spatula as a makeshift megaphone. Similarly, in religious discourse, an allegory in a sacred text may only be interpreted in its most traditional sense, overlooking potential modern or metaphorical meanings.

Impact: This bias can limit creativity and problem-solving abilities, whether it's in engineering a quick-fix solution to a technical glitch at work, or in interpreting and applying ancient religious texts to contemporary social issues.

Related Biases: Einstellung effect, where a person defaults to a solution that worked in the past rather than seeking a novel solution, is a kissing cousin of Functional Fixedness.

Countermeasures: Flex your creative muscles. Regularly challenging yourself to think of alternative uses for everyday objects can help. The next time you look at a coin, don't just see a monetary instrument, see also a potential screwdriver or a makeshift ruler.

Key Takeaways: Functional Fixedness is like having tunnel vision for object usage. It can limit your problem-solving and creative potential. So go ahead, take off those function-tinted glasses and see the world of possibilities that lies in your everyday objects.

References:

Duncker, K., & Lees, L. S. (1945). On problem-solving. Psychological Monographs, 58(5), i.

German, T. P., & Barrett, H. C. (2005). Functional fixedness in a technologically sparse culture. Psychological Science, 16(1), 1-5.

Fundamental Attribution Error

Definition: The Fundamental Attribution Error is a cognitive bias that prompts us to overestimate personality traits and underestimate situational influences when analyzing others' behavior.

Description: This bias gives the mind a free pass to slip into detective mode, analyzing people's actions as if they're clues to their character, rather than considering the context. Why did John spill his coffee? Because he's a clumsy oaf, not because someone jostled him.

Examples: For a religious spin, imagine someone who rarely attends church services. The Fundamental Attribution Error might lead you to conclude they lack faith, instead of considering factors like their work schedule or transportation access. In politics, if a politician makes an unpopular decision, it's easy to label them as self-serving, rather than considering the pressures and constraints they might be under.

Impact: This bias can distort how we perceive others, fueling stereotypes, misunderstanding, and conflict. It's especially problematic in polarized political environments, where decontextualized actions can become ammunition for character assassinations.

Related Biases: Actor-Observer Bias, where we attribute our own actions to the situation but others' actions to their dispositions, is like a sibling to the Fundamental Attribution Error.

Countermeasures: Embrace context. Before making judgments about others, consider their circumstances, pressures, and potential reasons for their behavior. Try walking a mile in their shoes, or at least, pondering the pebbles in them.

Key Takeaways: Remember, we live in a world shaped by circumstances and personalities. The Fundamental Attribution Error can skew our perception towards the latter. Recognizing this can lead to fairer, more empathetic assessments of others.

References:

Ross, L. (1977). The intuitive psychologist and his shortcomings: Distortions in the attribution process. In L. Berkowitz (Ed.), Advances in experimental social psychology (Vol. 10, pp. 173-220). Academic Press.

Gilbert, D. T., & Malone, P. S. (1995). The correspondence bias. Psychological Bulletin, 117(1), 21-38.

Gambler's Fallacy

Definition: Gambler's Fallacy, the mistaken belief that if an event occurred more frequently than normal during a past period, it's less likely to happen in the future (or vice versa).

Description: It's like believing that the universe has a "fairness quota" and if you've had a rough patch, you're due for some good luck. But alas, the universe isn't a gambler looking to even out the odds. Independent events, like coin flips or lottery draws, don't remember what happened before. They're historical amnesiacs, the past is as irrelevant to them as last year's Super Bowl is to a goldfish.

Examples: In politics, suppose a party has won three elections in a row. Some might think, "Surely, they can't win a fourth, right? The universe is due for a change!" In religion, imagine someone who's experienced a string of hardships. They might believe they're due for some divine good fortune, under the fallacy that the universe somehow balances out suffering and joy.

Impact: This bias can lead to irrational decisions in everything from gambling to predicting election outcomes to financial investing. You can lose money, credibility, and potentially a sense of reality when the universe refuses to "balance" things out.

Related Biases: Hot-hand Fallacy, the belief that a person who has experienced success has a higher chance of further success, is a gambler's fallacy cousin.

Countermeasures: Understanding probability and the independence of random events can help. Remember, the universe doesn't have an equity agenda, nor a scorecard.

Also, be cautious of patterns in random data; our brains are hardwired to see them, even when they're not there.

Key Takeaways: Remember, the Gambler's Fallacy is a sucker's bet. No matter how many times the coin landed on heads, the next flip still has a 50/50 chance. Don't expect the universe to even out your odds—it's got bigger fish to fry.

References:

Burns, B. D., & Corpus, B. (2004). Randomness and inductions from streaks: "Gambler's fallacy" versus "hot hand". Psychonomic Bulletin & Review, 11(1), 179-184.

Tune, G. S. (1964). Response preferences: A review of some relevant literature. Psychological Bulletin, 61(4), 286-302.

Generation Bias

Definition: Generation Bias is the tendency for individuals to perceive their own generation as superior to others, resulting in an unjustified preference for their own cohort.

Description: Every generation has its pet peeves about the other. Baby Boomers often grumble about Millennials and their avocado toast obsession, while Gen Zers may bemoan the perceived rigidity of older generations. It's like an age-based turf war, but with fewer leather jackets and more social media platforms.

Examples: Politically, this can be seen in voting patterns where certain generations may favor policies that mostly benefit their own cohort. In religious contexts, older generations might view their traditional religious practices as superior, dismissing younger generation's flexible spiritual pursuits as 'fads.'

Impact: This bias can lead to intergenerational conflict, impede progress, and reinforce stereotypes. It might also hamper constructive dialogue and mutual understanding between generations, creating rifts in societal, political, and familial contexts.

Related Biases: Ingroup Bias, where people favor their own group over others, is a close relative of the Generation Bias.

Countermeasures: Cultivate awareness of this bias and actively seek to understand the perspectives of other generations. Encourage intergenerational communication and collaboration. Remember, every generation has its strengths and weaknesses, not unlike different flavors of ice cream—some might prefer classic vanilla while others rave about salted caramel.

Key Takeaways: Generation Bias can cloud our judgment and impede collaboration. Recognizing the value in each generation, like appreciating different ice cream flavors, can lead to a more harmonious society. After all, one generation's avocado toast is another generation's deviled eggs.

References:

North, M. S., & Fiske, S. T. (2012). An inconvenienced youth? Ageism and its potential intergenerational roots. Psychological Bulletin, 138(5), 982–997.

Putney, N. M., & Bengtson, V. L. (2003). Intergenerational relations in changing times. In G. A. Kimble & M. Wertheimer (Eds.), Great minds reflect on the good life: Aging and the life course (pp. 149–164). American Psychological Association.

Google Effect

Definition: The Google Effect, also known as digital amnesia, is a cognitive bias where individuals tend to forget information that can be easily found using internet search engines such as Google.

Description: It's like delegating the task of remembering to the internet, making your brain the boss that doesn't need to know all the details as long as it knows whom to ask. The brain, after all, has unionized and is constantly negotiating for less storage work.

Examples: Consider political debates online. Instead of internalizing facts and data about policies, we rely on Google to supply information on demand. In religious discourse, individuals may not remember specific scripture because they know they can easily look it up online.

Impact: The Google Effect might impact our ability to retain and recall information, potentially altering how our memory works. Our reliance on the internet could weaken cognitive functions, or it might simply be a form of cognitive offloading, freeing our brains for other tasks. We're like the proud parents of a not-so-bright child, boasting that our brain may not remember much, but at least it knows where to find information.

Related Biases: This is closely related to the transactive memory system, where we rely on others to remember information for us.

Countermeasures: Try to balance the use of digital tools with memory-enhancing activities. Quiz yourself before resorting to search engines, and practice mindfulness to improve memory retention. After all, a well-exercised brain could come in handy during that inevitable zombie apocalypse when Google might not be available.

Key Takeaways: While the Google Effect shows how reliant we've become on the internet for information, it's essential to strike a balance. The human brain is still the best search engine you have, and it doesn't even require a Wi-Fi connection.

References:

Sparrow, B., Liu, J., & Wegner, D. M. (2011). Google Effects on Memory: Cognitive Consequences of Having Information at Our Fingertips. Science, 333(6043), 776-778.

Wegner, D. M., & Ward, A. F. (2013). The Internet Has Become the External Hard Drive for Our Memories. Scientific American, 309(6), 58-61.

Group Attribution Error

Definition: Group Attribution Error is the cognitive bias that leads us to believe the characteristics of an individual group member reflect the group as a whole.

Description: It's our brain's shortcut, saying, "Why waste energy understanding complex group dynamics when I can just assign characteristics willy-nilly?" It's the mental equivalent of judging a book by its cover, except it's a book club, and we're basing our judgment on one loud-mouthed member.

Examples: In politics, we might perceive all members of a particular party as having the same beliefs based on the views of a few prominent figures. Similarly, in religious contexts, we might attribute the actions or beliefs of a few individuals to the entire congregation or faith.

Impact: Group Attribution Error can lead to stereotypes, prejudice, and unfair judgments. It paints an inaccurate picture of group diversity and individual uniqueness. It's like assuming all kittens are mischievous because one clawed your favorite couch.

Related Biases: This bias is closely related to the Fundamental Attribution Error and Stereotyping.

Countermeasures: Challenge your assumptions by seeking information about individuals, not just groups. Avoid hasty generalizations and remind yourself of human complexity. Just as all your toes aren't the same length, all group members don't share the same characteristics.

Key Takeaways: While our brains might love the shortcut of Group Attribution Error, it can lead to misconceptions and unfair judgments. Always remember to honor individuality and diversity.

References:

Tajfel, H. (1982). Social psychology of intergroup relations. Annual review of psychology, 33(1), 1-39.

Hewstone, M. (1990). The 'ultimate attribution error'? A review of the literature on intergroup causal attribution. European Journal of Social Psychology, 20(4), 311-335.

Groupthink

Definition: Groupthink is a cognitive bias where the desire for group consensus overrides people's common sense desire to present alternatives, critique a position, or express an unpopular opinion.

Description: Imagine groupthink as an overenthusiastic cheerleader, drowning out any dissenting voices to keep the team spirit high. It happens when the fear of rocking the boat triumphs over rational decision-making. Remember, not every boat needs to be rocked, but some could use a good capsizing.

Examples: In politics, a political party might continue to support a policy not because it's effective but because it's the party line. In religion, a congregation may continue following a doctrine because it's tradition, even if some members harbor doubts.

Impact: Groupthink can lead to poor decision-making, suppression of individual creativity, and even catastrophic outcomes. It's a bit like letting the blind lead the blind in a direction nobody really wants to go, but everyone is too polite to say so.

Related Biases: This bias is closely related to Conformity, Polarization, and Bandwagon Effect.

Countermeasures: Promote open dialogue, create a safe environment for dissenting opinions, and remember that a disagreement is not an argument – it's a sign of a healthy, functioning group. Sometimes the lone voice crying out in the wilderness is the one you need to hear.

Key Takeaways: Groupthink suppresses individual thought in favor of consensus, often leading to poor decisions. Encourage diverse thinking and remember that it's okay to break away from the choir and hit your own note.

References:

Janis, I. L. (1972). Victims of groupthink: A psychological study of foreign-policy decisions and fiascoes. Boston: Houghton Mifflin.

Turner, M. E., & Pratkanis, A. R. (1998). Twenty-five years of groupthink theory and research: Lessons from the evaluation of a theory. Organizational Behavior and Human Decision Processes, 73(2-3), 105-115.

Hard-Easy Effect

Definition: The Hard-Easy effect is a cognitive bias where people tend to overestimate their ability to predict the outcome of hard tasks and underestimate their ability to predict the outcome of easy tasks.

Description: We can compare the Hard-Easy Effect to a cat attempting to leap across a wide chasm or snuggle into a tiny box. The feline might boldly attempt the chasm (and fail), while doubting its ability to fit into the box (when it obviously can). The kitty has fallen prey to the Hard-Easy effect, just like humans do every day.

Examples: In politics, a pundit might overestimate their ability to predict the outcome of a complex political situation (a "hard" task) while underestimating their ability to predict the outcome of a straightforward policy vote (an "easy" task). In religious contexts, people might overstate their ability to interpret a complex theological doctrine while underrating their understanding of simpler beliefs.

Impact: The Hard-Easy effect can lead to misguided decisions, poor predictions, and a misunderstanding of our own skills and abilities. It's like bringing a toothpick to a sword fight and a rocket launcher to a knife fight - the former is a disaster waiting to happen and the latter is overkill.

Related Biases: Overconfidence Bias, Dunning-Kruger effect, and Optimism Bias are all siblings in the family of misguided self-evaluation biases.

Countermeasures: Regular self-evaluation and feedback can help keep our predictive abilities in check. It's also beneficial to remember that a superhero costume doesn't grant superpowers, no matter how good you look wearing it.

Key Takeaways: We often misjudge our ability to predict outcomes, particularly when it comes to hard and easy tasks. Regular reality checks can help keep us grounded in our true predictive capabilities.

References:

Keren, G. (1980). The "Challenging Items Effect" and the differential difficulty model. Acta Psychologica, 45(2), 109-119.

Lichtenstein, S., & Fischhoff, B. (1977). Do those who know more also know more about how much they know?. Organizational Behavior and Human Performance, 20(2), 159-183.

Halo Effect

Definition: The Halo Effect is a cognitive bias where positive impressions in one area influence our opinions in other areas. If we see an angel's halo, we tend to assume the rest of the being is angelic too.

Description: The Halo Effect is like having rose-colored glasses permanently glued to your face. Once we form a positive impression of someone or something, this "halo" can obscure any flaws, making everything seem just peachy.

Examples: In politics, a charismatic leader who gives great speeches (that positive halo) may lead us to overlook their lackluster policy decisions. In religion, a popular figure who is well-spoken and kind can cloud our judgment about their less than saintly behaviors. In social issues, a well-liked celebrity who advocates for a cause may make us automatically support that cause, without considering its complexities.

Impact: The Halo Effect can lead to misjudgments, oversights, and in some cases, putting faith or trust in undeserving individuals or causes. It's like eating a beautifully decorated cake only to find out it's made of cardboard—quite the letdown.

Related Biases: The Halo Effect is related to the Horn Effect (the devil's in the details) and the Similar-to-me Effect (like looking in a mirror).

Countermeasures: To counter the Halo Effect, practice critical thinking and independent evaluation of different attributes. Don't judge a book by its cover or a person by their halo.

Key Takeaways: Our perceptions can be heavily influenced by positive impressions, leading to potential errors in judgment. Maintain a healthy skepticism and evaluate individual aspects independently to mitigate the halo effect.

References:

Nisbett, R. E., & Wilson, T. D. (1977). The halo effect: Evidence for unconscious alteration of judgments. Journal of Personality and Social Psychology, 35(4), 250–256.

Thorndike, E. L. (1920). A constant error in psychological ratings. Journal of Applied Psychology, 4(1), 25–29.

Hawthorne Effect

Definition: The Hawthorne Effect is a phenomenon where individuals modify or improve their behavior when they know they're being observed. It's the grown-up version of kids suddenly cleaning up their rooms when they know mom and dad are watching.

Description: The Hawthorne Effect happens when you're on your best behavior because you know someone is watching. It's not just about impressing the boss or the researcher, though. The effect is named after a series of studies at the Western Electric Hawthorne Works in Chicago, where researchers found that any change in work conditions temporarily increased productivity. It was less about the changes and more about feeling watched and important.

Examples: Politicians might act more 'statesmanlike' when they're in the public eye (and then perhaps less so behind closed doors). Religious leaders may be on their best behavior when congregants are present. On social issues, public figures might advocate passionately when the cameras are rolling, but their off-camera commitment might not be as robust.

Impact: The Hawthorne Effect can lead to false perceptions about people's typical behavior, resulting in inflated expectations or misplaced trust. It's like thinking every day is a party because you always see your neighbors during their weekend barbecues.

Related Biases: Social Desirability Bias (the need to be seen in a positive light) and Observer-expectancy Effect (where the observer inadvertently influences the participants).

Countermeasures: To combat the Hawthorne Effect, consider anonymous observations or measure performance over an extended period. And remember, not everyone is as well-behaved as they seem when they know you're watching.

Key Takeaways: Observation can alter behavior, often for the better, but this doesn't necessarily reflect a person's true character or commitment. Keep the Hawthorne Effect in mind when interpreting the actions of those under observation.

References:

Adair, J. G. (1984). The Hawthorne effect: A reconsideration of the methodological artifact. Journal of Applied Psychology, 69(2), 334–345.

Gillespie, R. (1991). Manufacturing knowledge: A history of the Hawthorne experiments. Cambridge University Press.

Herd Instinct

Definition: Herd Instinct, often referred to as "going along with the crowd", is the propensity for individuals to adopt the behaviors, attitudes, and actions of the majority of people within a group. It's the psychological driving force behind why you might order what everyone else is getting at a restaurant, even if you've never heard of quinoa.

Description: This cognitive bias arises from our inherent desire to fit in, avoid conflict, and feel secure within a group setting. We're social creatures by nature, and sticking with the herd has been a survival strategy for eons. But just because it helped our ancestors avoid being lunch doesn't mean it always serves us well today.

Examples: In politics, herd instinct might result in party-line voting, even when constituents don't agree with all the party's stances. In religion, followers may adopt the majority's beliefs without personal exploration or understanding. On social issues, people might publicly express popular opinions to avoid standing out or facing criticism, even if privately they disagree.

Impact: The Herd Instinct can stifle individual thought, encourage conformity, and perpetuate unexamined beliefs. It's also responsible for the spread of misinformation, since the 'popular' view isn't always the correct or informed one. Ever heard of the world being flat?

Related Biases: Groupthink (suppressing dissenting opinions to maintain harmony) and Bandwagon Effect (adopting beliefs because of their growing popularity).

Countermeasures: Develop critical thinking skills, seek out diverse perspectives, and remember it's okay to go against the herd. In fact, some of the most transformative ideas in history have come from those willing to stray from the pack.

Key Takeaways: While sticking with the herd might feel safe, it's important to think for yourself, question commonly accepted beliefs, and feel secure in your individual decisions. You don't have to eat quinoa if you really don't want to.

References:

Raafat, R. M., Chater, N., & Frith, C. (2009). Herding in humans. Trends in Cognitive Sciences, 13(10), 420–428.

Le Bon, G. (2002). The Crowd: A Study of the Popular Mind. Dover Publications. (Original work published 1895)

Hindsight Bias

Definition: Hindsight bias is a cognitive quirk that causes us to believe, after an event has occurred, that we had accurately predicted or expected it beforehand. Think of it as the "I-knew-it-all-along" effect or the mental equivalent of claiming you always knew your ex was a werewolf...after you see them sprouting fur under a full moon.

Description: The human brain is an impressive prediction machine, but it has its hiccups. One of them is this sneaky little bias that makes us retroactively award ourselves psychic abilities. It stems from our need to find order and predictability in the world, which, to be fair, is a little chaotic and full of surprise werewolves.

Examples: In religion, people may retroactively fit prophecies to events ("The prophecy did mention a great calamity, didn't it?"). In politics, supporters of a policy might insist they foresaw its success, while critics claim they predicted its failure. And who hasn't said, "I knew that was going to happen!" when a social situation goes exactly as disastrously as it did in your anxiety dreams?

Impact: Hindsight bias can lead to overconfidence in our predictive abilities and hamper learning from past mistakes. It can even impact legal and judicial decisions, leading to unfair expectations and judgments.

Related Biases: Confirmation bias (seeking out information that confirms pre-existing beliefs), outcome bias (judging a decision based on its outcome rather than its quality at the time).

Countermeasures: Regular reality checks can help. Remind yourself that the world is inherently unpredictable, and the future is not a rerun of the past. Keeping a decision journal, where you record your expectations and reasons for making decisions, can also be useful to check your biases.

Key Takeaways: The hindsight bias paints a rosy (and misleading) picture of our predictive prowess. Awareness of this bias can help you keep your self-perception in check and promote more rational decision-making.

References:

Fischhoff, B. (1975). Hindsight ≠ foresight: The effect of outcome knowledge on judgment under uncertainty. Journal of Experimental Psychology: Human Perception and Performance, 1(3), 288–299.

Roese, N. J., & Vohs, K. D. (2012). Hindsight Bias. Perspectives on Psychological Science, 7(5), 411–426.

Hoarding

Definition: Hoarding refers to the excessive acquisition of items, even if these items are worthless or have no space to store them, coupled with a persistent difficulty in parting with them. It's like the human version of a squirrel stockpiling nuts for a winter that never comes.

Description: Hoarding isn't just about being overly fond of stuff. It's a complex behavioral pattern often linked to a range of other psychological disorders such as OCD, depression, and anxiety disorders. It's about more than just a cluttered garage or a well-stocked pantry. Hoarding often has deep psychological roots that turn "one man's trash" into "this man's impossible-to-discards."

Examples: A classic case might be your uncle who has every newspaper since 1974, just in case he might need a coupon that expired in 1975. Politically, think of policy makers who keep unnecessary, outdated laws on the books because they can't bear to part with them.

Impact: Hoarding can have significant impacts on an individual's life, including social isolation, fire hazards, and other health risks. It can also have societal impacts, as in times of crisis, hoarding behavior can exacerbate shortages and increase panic.

Related Biases: The Endowment Effect (overvaluing something merely because we own it), Loss Aversion (the pain of losing is psychologically twice as powerful as the pleasure of gaining).

Countermeasures: Cognitive-behavioral therapy (CBT) is currently the most effective treatment for hoarding disorder.

Techniques used in CBT for hoarding include categorization and organization skills, non-acquisition and discarding skills training, exposure to discarding and not acquiring, and cognitive restructuring around problematic beliefs about possessions.

Key Takeaways: Hoarding is a complex issue with psychological roots. Being aware of this bias can help in understanding the need for professional help to address it.

References:

Tolin, D. F., Frost, R. O., Steketee, G., Gray, K. D., & Fitch, K. E. (2008). The economic and social burden of compulsive hoarding. Psychiatry Research, 160(2), 200–211.

Steketee, G., & Frost, R. O. (2003). Compulsive Hoarding: Current Status of the Research. Clinical Psychology Review, 23(7), 905–927.

Horns Effect

Definition: The Horns Effect is a cognitive bias that prompts you to negatively perceive someone based on a single unfavorable characteristic. It's like judging a book by its worst chapter and not even considering the redeeming qualities of the remaining chapters.

Description: It's the doppelganger of the Halo Effect, but with a devilish twist. This bias causes you to allow one negative trait to overshadow all other aspects of a person. Essentially, we paint an individual with a broad negative brush because of one aspect we don't particularly like.

Examples: In politics, a candidate might propose a groundbreaking policy, but all voters remember is that one scandal they were part of years ago. In the realm of religion, a church member might be known for their devotion and community service, but let them sit in your favorite pew one time...

Impact: The Horns Effect can lead to misjudgment and unfair treatment. In the workplace, it could lead to discrimination and biased decision-making. In politics, it might cause voters to overlook competent candidates due to insignificant or irrelevant missteps.

Related Biases: The Halo Effect (the bias's more angelic twin), First Impression Bias, and Confirmation Bias.

Countermeasures: Encourage objective thinking and evaluations based on a full range of evidence, not just a single attribute. In organizations, structured and standard assessments can help reduce the influence of the Horns Effect.

Key Takeaways: The Horns Effect is a cognitive bias that can lead us to unfairly judge others based on a single negative characteristic. To counter this, it's crucial to look at the bigger picture and not let one negative aspect cloud our entire perception.

References:

Nisbett, R. E., & Wilson, T. D. (1977). The halo effect: Evidence for unconscious alteration of judgments. Journal of Personality and Social Psychology, 35(4), 250-256.

Kaplan, R. M. (1978). Is beauty talent? Sex interaction in the attractiveness halo effect. Sex Roles, 4(2), 195-204.

Hot-Hand Fallacy

Definition: The Hot-Hand Fallacy is the cognitive bias that leads people to presume a streak of successful outcomes will continue in the future, simply due to the recent successes.

Description: This is essentially Lady Luck's evil twin, doling out disappointment instead of fortune. The term originates from basketball, where a player who's made a few shots in a row is seen as having a "hot hand," even though each shot is an independent event and previous success doesn't predict future outcomes. Our brains, however, aren't great at processing randomness and probabilities, so we end up believing in the gambler's fallacy - a twist of which is the hot-hand fallacy.

Examples: In politics, a candidate who's won a series of debates might be seen as invincible, even though they could trip over their tongue at any moment. In religion, a pastor might deliver a few popular sermons in a row and suddenly, he's the next Billy Graham. But wait until he tries to walk on water...

Impact: The Hot-Hand Fallacy can lead to overconfidence, poor decision-making, and potential losses, as individuals may take greater risks based on the incorrect belief that their luck will hold.

Related Biases: Gambler's Fallacy, Illusion of Control, and Overconfidence Bias.

Countermeasures: Understanding randomness and the independent nature of events can help mitigate this bias. Education about probabilities and statistics can be beneficial.

Key Takeaways: The Hot-Hand Fallacy is a cognitive bias that leads us to believe success will continue simply because it has in the past. To counter this, it's essential to understand the independent nature of events and not let success inflate our expectations.

References:

Gilovich, T., Vallone, R., & Tversky, A. (1985). The hot hand in basketball: On the misperception of random sequences. Cognitive Psychology, 17(3), 295-314.

Burns, B. D., & Corpus, B. (2004). Randomness and inductions from streaks: "Gambler's fallacy" versus "hot hand". Psychonomic Bulletin & Review, 11(1), 179-184.

Hot-Cold Empathy Gap

Definition: The Hot-Cold Empathy Gap is the cognitive bias that leads people to underestimate the influence of emotional states (hot states such as hunger, anger, or fatigue) on their own and others' decision-making and behavior when they are in a neutral state (cold state).

Description: This bias has us playing a little game of emotional meteorology—forecasting others' reactions based on our current internal climate, rather than theirs. Picture yourself enjoying a cool ice cream on a hot summer day, and suddenly presuming everyone else in the world couldn't possibly be in a bad mood. See, easy to fall into, isn't it?

Examples: In politics, leaders often make promises in "cold" states without realizing how much their decisions will change when they are in the "hot" state of political pressure. In social issues, a person in a comfortable and secure environment might underestimate the desperation of someone in a crisis situation, leading to a lack of empathy or inappropriate policy decisions.

Impact: The Hot-Cold Empathy Gap can lead to misunderstandings, miscommunication, and poor decision-making. It can also lead to a lack of empathy and an inability to accurately predict our future behavior.

Related Biases: Projection Bias, Affective Forecasting Errors.

Countermeasures: Increasing our awareness of this bias and trying to consciously consider the emotional state of others can help. Role-playing or empathy exercises may also be beneficial.

Key Takeaways: The Hot-Cold Empathy Gap is a cognitive bias that causes us to underestimate the influence of emotional states on decision-making. Recognizing this can improve communication, decision-making, and empathy.

References:

Loewenstein, G. (2005). Hot-cold empathy gaps and medical decision making. Health Psychology, 24(4S), S49.

Van Boven, L., & Loewenstein, G. (2003). Social projection of transient drive states. Personality and Social Psychology Bulletin, 29(9), 1159-1168.

Hyperbolic Discounting

Definition: Hyperbolic Discounting is the cognitive bias where people choose smaller-sooner rewards over larger-later rewards, demonstrating an inconsistent valuation over time. It's like a financial advisor for instant gratification.

Description: In layman's terms, we humans have a "want it now" button that's often set to auto-pilot. While we know in theory that "patience is a virtue," we're also biologically wired to seize the day—or the donut—whatever the case may be.

Examples: You might see this bias in play during political campaigns where promises of immediate tax cuts or immediate benefits often sway voters, despite the long-term consequences. Or in religion, where the promise of instant salvation often overrides the idea of lifelong spiritual discipline. Or in social issues, where short-term solutions often get more support than long-term sustainable policies.

Impact: Hyperbolic Discounting can lead to impulsive decision-making and short-term thinking, which can negatively impact long-term goals and plans. It can also contribute to procrastination, unhealthy behaviors, and financial mismanagement.

Related Biases: Present Bias, Instant Gratification Bias, Time Inconsistency.

Countermeasures: Increasing self-awareness and practicing mindfulness can help one pause before making impulsive decisions. In addition, goal-setting, planning, and imagining the future can help mitigate the influence of this bias.

Key Takeaways: Hyperbolic Discounting reflects our tendency to favor immediate rewards over future ones. Recognizing this bias can lead to more balanced decision-making and better long-term planning.

References:

Frederick, S., Loewenstein, G., & O'donoghue, T. (2002). Time discounting and time preference: A critical review. Journal of Economic Literature, 40(2), 351-401.

Laibson, D. (1997). Golden eggs and hyperbolic discounting. Quarterly Journal of Economics, 112(2), 443-478.

Illusion of Asymmetric Insight

Definition: The Illusion of Asymmetric Insight is a cognitive bias where individuals believe they understand others better than others understand them. It's the cognitive version of "I know you are, but what am I?"

Description: This delightful mental tic convinces us that we possess an illuminating, private window into the hearts and minds of others, while they're merely gazing at our opaque exterior. Essentially, we're all walking around thinking we're the Sherlock Holmes of human interaction.

Examples: This bias often finds a home in the political realm, where individuals of one political group are convinced they understand the motives and flaws of the opposing group better than they are understood. Similarly, in religious discussions, believers may feel they understand non-believers better than they are understood. This bias also thrives in social issues, where one group claims superior understanding of another group's experiences and perspectives.

Impact: The illusion of asymmetric insight can lead to misunderstanding, conflict, and a lack of productive communication. It fuels stereotypes, hinders empathetic understanding, and promotes an 'us versus them' mentality.

Related Biases: False Consensus Effect, Fundamental Attribution Error, Naïve Cynicism.

Countermeasures: Self-awareness, empathy, and open dialogue can help counteract this bias. A good dose of humility and the recognition that understanding others (and even ourselves) is a complex task can go a long way in promoting more balanced perceptions.

Key Takeaways: The Illusion of Asymmetric Insight can lead us to overestimate our understanding of others while underestimating their understanding of us. Recognizing this bias can lead to more empathetic and effective communication.

References:

Pronin, E., Lin, D. Y., & Ross, L. (2002). The Bias Blind Spot: Perceptions of Bias in Self Versus Others. Personality and Social Psychology Bulletin, 28(3), 369–381.

Kruger, J., & Gilovich, T. (1999). "Naive cynicism" in everyday theories of responsibility assessment: On biased assumptions of bias. Journal of Personality and Social Psychology, 76(5), 743–753.

Illusion of Attention

Definition: The Illusion of Attention refers to the cognitive bias where individuals overestimate the amount of attention they pay to their environment or to a given task. It's the reason why we confidently assure our partners that "Yes, dear, I was listening," when we, in fact, were daydreaming about dinner.

Description: Our brains have this charming knack for making us believe we're the star students of life's classroom, attentively noting every detail of our surroundings. Spoiler alert: we're not. The truth is, our attention is a limited resource, and we are notoriously bad at multitasking.

Examples: For example, in the realm of politics, a person may believe they are well-informed about a particular topic, having read a few headlines and tweets, while in reality, their understanding is superficial. This can also apply to religious debates or discussions around social issues where people rely on snippets of information without delving into the complexities.

Impact: The Illusion of Attention can lead to misinformation, miscommunication, and false confidence in our knowledge or awareness. This can affect decision-making, judgement, and perception, and can contribute to a lack of in-depth understanding and empathy in societal matters.

Related Biases: Attentional Bias, Overconfidence Effect, Dunning-Kruger Effect.

Countermeasures: Mindfulness and focused attention can be effective countermeasures. Slow down, concentrate on one thing at a time, and make an effort to engage more deeply with information and experiences.

Key Takeaways: The Illusion of Attention convinces us we're more observant and attentive than we really are. Mindfulness and focused attention can help combat this bias, leading to a more accurate understanding and effective communication.

References:

Simons, D. J., & Chabris, C. F. (1999). Gorillas in our midst: Sustained inattentional blindness for dynamic events. Perception, 28(9), 1059-1074.

Kahneman, D. (2011). Thinking, Fast and Slow. New York, NY: Farrar, Straus and Giroux.

Illusion of Control

Definition: The Illusion of Control is the tendency for individuals to believe that they can control or influence outcomes that are, for the most part, beyond their control. It's like buying a lottery ticket and believing you'll win because you chose your "lucky" numbers.

Description: It's human nature to yearn for control over our circumstances, and our brains kindly oblige by serving up this comforting illusion. Psychologically, it's rooted in our innate desire to assert control over our environment and lives. However, this can often lead us to overlook the randomness of many events.

Examples: Within politics, this can occur when a person believes that sharing their political opinions on social media will substantially sway public sentiment. In the realm of religion, it might be a person believing that their prayers or rituals directly influence outcomes, such as curing an illness or winning a sports game.

Impact: This bias can lead to overconfidence, poor decision-making, and an avoidance of personal responsibility when things go south. It can also cloud our perception of reality and reduce our ability to assess situations objectively.

Related Biases: Overconfidence Bias, Gambler's Fallacy, and Optimism Bias.

Countermeasures: It can be helpful to regularly remind ourselves of the limits of our control, and to embrace uncertainty as an inherent part of life.

Key Takeaways: The Illusion of Control leads us to overestimate our influence over external events. By understanding the limits of our control and embracing uncertainty, we can make more informed decisions and reduce the potential for disappointment when outcomes don't align with our expectations.

References:

Langer, E. J. (1975). The illusion of control. Journal of Personality and Social Psychology, 32(2), 311-328.

Presson, P. K., & Benassi, V. A. (1996). Illusion of control: A meta-analytic review. Journal of Social Behavior and Personality, 11(3), 493-510.

Illusion of External Agency

Definition: The Illusion of External Agency refers to the cognitive bias in which individuals attribute their thoughts, feelings, or behaviors to some outside force, be it fate, a deity, or possibly a weather-controlling machine in someone's basement.

Description: Rather than accept personal responsibility or randomness, our brain, the ultimate screenwriter, prefers to draft a more engaging narrative in which external agents pull the strings. It's a form of psychological defense mechanism, protecting us from the harsh realities of our limitations or the random chaos of the universe.

Examples: For instance, consider a politician who, despite a mountain of evidence against them, maintains the belief that a nebulous "deep state" is responsible for their misfortunes. In religion, we might see the Illusion of External Agency at play when people attribute natural disasters to divine wrath rather than environmental factors.

Impact: This cognitive bias can hinder personal growth and accountability, as it shifts blame for one's actions and consequences onto imaginary external factors. It can also lead to a sense of learned helplessness, as individuals feel powerless to change their circumstances.

Related Biases: Fundamental Attribution Error, Self-Serving Bias, and Just-world Bias.

Countermeasures: Regular introspection, and reality checks can help mitigate this bias. It's important to understand that not everything is within our control, but simultaneously, not everything is beyond our control either.

Key Takeaways: The Illusion of External Agency can create a convenient scapegoat for our own actions or the randomness of life, but it can also inhibit personal growth and foster a sense of helplessness. Balancing personal accountability with the acceptance of external factors is crucial for a more objective understanding of our world.

References:

Wegner, D. M., Sparrow, B., & Winerman, L. (2004). Vicarious agency: Experiencing control over the movements of others. Journal of Personality and Social Psychology, 86(6), 838–848.

Desantis, A., Roussel, C., & Waszak, F. (2011). On the influence of causal beliefs on the feeling of agency. Consciousness and cognition, 20(4), 1211-1220.

Illusion of Explanatory Depth

Definition: The Illusion of Explanatory Depth (IoED) is the charming yet misleading belief that we understand things far better than we really do. It's like a deceivingly shallow puddle that you think you can skip over, only to end up soaked up to your knees.

Description: We humans tend to overestimate our understanding of complex phenomena, from everyday objects to intricate socio-political issues. It's not until we're asked to explain these phenomena in detail that we recognize the limit of our knowledge. This bias emerges from the human tendency to simplify the world to make it manageable, resulting in a cognitive equivalent of a child's drawing, not a photorealistic portrait.

Examples: A prime example would be your Uncle Bob at family gatherings, confidently explaining the inner workings of the global economy or climate change while dolloping mashed potatoes on his plate. It becomes clear he might not be the expert when you ask him to elaborate, and he quickly changes the topic to his golf handicap.

Impact: IoED can be a roadblock to learning and progress. It may lead to overconfidence, poor decision-making, and even conflicts when people staunchly defend their superficial understanding of complex issues.

Related Biases: Dunning-Kruger Effect, Overconfidence Bias, and Curse of Knowledge.

Countermeasures: A good defense against IoED is humility and curiosity. Recognize that it's okay not to know everything, and nurture a love for learning. Actively seek out information and don't shy away from asking questions.

Key Takeaways: Our minds may make us feel like mini Einsteins, but in reality, our knowledge often has more holes than Swiss cheese. Embrace this fact, and you'll be more open to learning and less likely to make hasty, uninformed decisions.

References:

Rozenblit, L., & Keil, F. (2002). The misunderstood limits of folk science: an illusion of explanatory depth. Cognitive science, 26(5), 521-562.

Alter, A. L., Oppenheimer, D. M., & Zemla, J. C. (2010). Missing the trees for the forest: A construal level account of the illusion of explanatory depth. Journal of personality and social psychology, 99(3), 436–451.

Illusion of Knowledge

Definition: The Illusion of Knowledge is a cognitive bias where people believe they understand more than they do. It's akin to seeing the headline and believing you've read the whole article - spoiler alert, you haven't!

Description: This bias causes us to overestimate our knowledge or understanding of a subject based on superficial information or cursory exposure. The complexity and depth of the subject are often ignored or misunderstood, giving us an illusion of expertise that's as flimsy as a house of cards in a tornado.

Examples: A prime example could be a voter who reads a few headlines or social media posts about a particular policy and considers themselves an expert, ready to take on anyone in a debate. Alas, if they were asked to describe the specifics of the policy, they might find themselves stumbling more than a toddler in oversize shoes.

Impact: The Illusion of Knowledge can lead to unproductive arguments, poor decisions, and misinformation spreading like wildfire. It can particularly fuel polarization in political, religious, and social matters when individuals' perceived knowledge outpaces their actual understanding.

Related Biases: Dunning-Kruger Effect, Overconfidence Bias, and Illusion of Explanatory Depth.

Countermeasures: The antidote to this illusion is humility, an eagerness to learn, and a healthy skepticism of one's knowledge. Engaging in active learning, seeking diverse sources of information, and inviting differing perspectives can help.

Key Takeaways: Just like salt enhances a dish, a sprinkle of knowledge can spice up a conversation. However, mistaking a sprinkle for a full meal may leave you with an empty stomach and a lot of explaining to do.

References:

Friedman, D., & Miyake, A. (2004). The relations among inhibition and interference control functions: a latent-variable analysis. Journal of experimental psychology: General, 133(1), 101.

Dunning, D., Johnson, K., Ehrlinger, J., & Kruger, J. (2003). Why people fail to recognize their own incompetence. Current directions in psychological science, 12(3), 83-87.

Illusion of Skill

Definition: The Illusion of Skill refers to the cognitive bias where individuals overestimate their abilities or skills, often attributing success to skill rather than luck or chance. It's like thinking you're a virtuoso singer because you can nail "Twinkle Twinkle Little Star".

Description: This illusion tends to rear its overly confident head when outcomes are influenced by factors beyond our control, yet we still believe our skill was the star of the show. Like thinking we controlled the weather because it didn't rain on our picnic - news flash, you're not a weather deity.

Examples: Imagine a political pundit who correctly predicts the outcome of one election and suddenly thinks they have a Midas touch for political prognostication. They may overlook factors such as demographic trends or public sentiment, attributing their successful prediction to their "expertise" alone.

Impact: The Illusion of Skill can lead to poor decision-making, overconfidence, and a failure to accurately assess risk. It can cause significant misjudgments in areas such as politics or social issues where the complexity and unpredictable nature of human behavior often defy simplistic analysis.

Related Biases: Overconfidence Bias, Illusion of Control, and Dunning-Kruger Effect.

Countermeasures: To combat this illusion, engage in critical self-reflection, seek feedback, and cultivate an understanding of the role of chance and external factors in outcomes. Stay humble, my friend.

Key Takeaways: The Illusion of Skill is a reminder that not all that glitters is gold, and not all success is down to skill. Sometimes, you're just lucky, and that's okay. Just don't go betting the farm on it.

References:

Kruger, J., & Dunning, D. (1999). Unskilled and unaware of it: How difficulties in recognizing one's own incompetence lead to inflated self-assessments. Journal of Personality and Social Psychology, 77(6), 1121–1134.

Barber, B. M., & Odean, T. (2001). Boys will be boys: Gender, overconfidence, and common stock investment. Quarterly Journal of Economics, 116(1), 261–292.

Illusion of Superiority

Definition: The Illusion of Superiority, a close relative of narcissism, is a cognitive bias that causes people to overestimate their own abilities and qualities compared to others. It's the psychological version of "my dog is better than all the other dogs in the dog park."

Description: This bias is like wearing rose-tinted glasses when looking at oneself, but switching to regular lenses for everyone else. Not only does it inflate one's self-perception, but it can also foster a sense of entitlement, as if the universe's sole purpose is to serve one's whims.

Examples: In politics, a leader may believe their country is the "greatest nation on earth" simply by virtue of their own leadership. Or in religion, followers might believe their faith to be the "one true faith," disregarding the legitimacy of other faiths.

Impact: This illusion can lead to blind spots in our self-awareness, hinder personal growth, create conflict, and contribute to social divides. Believing we're always the smartest person in the room might win us a few points in a game of Trivial Pursuit, but it's not a great strategy for winning friends or influencing people.

Related Biases: Overconfidence Bias, Illusion of Control, and Dunning-Kruger Effect.

Countermeasures: Reality checks are crucial. Seek feedback, cultivate empathy, and maintain a learner's mentality. Try to see the world from others' perspectives and always question your assumptions. And remember, every dog owner thinks their dog is the best.

Key Takeaways: The Illusion of Superiority might feel good in the short term, but it won't win you any popularity contests. Recognizing this bias and striving for a more balanced self-perception can lead to better relationships, more effective decision-making, and a healthier dose of humility.

References:

Dunning, D., Johnson, K., Ehrlinger, J., & Kruger, J. (2003). Why people fail to recognize their own incompetence. Current Directions in Psychological Science, 12(3), 83-87.

Kruger, J., & Dunning, D. (1999). Unskilled and unaware of it: How difficulties in recognizing one's own incompetence lead to inflated self-assessments. Journal of Personality and Social Psychology, 77(6), 1121-1134.

Illusion of Transparency

Definition: The Illusion of Transparency is a cognitive bias that leads people to believe their thoughts, feelings, and emotions are more apparent to others than they actually are. It's like believing you're a walking, talking glass house when you're more like a brick building with heavily tinted windows.

Description: This bias can have you convinced that everyone around you has suddenly developed telepathic abilities. It may manifest when you blush after a gaffe and think everyone noticed, or when you expect your partner to intuit your desire for Thai food because you've been "giving hints" all evening.

Examples: In politics, a leader may assume their intentions are clear to constituents without explicit communication. Similarly, religious leaders might believe their congregation understands the deeper implications of a sermon when, in fact, they might only be scratching the surface.

Impact: This illusion can lead to miscommunication and misunderstandings. It can foster resentment when expectations go unmet and could contribute to social and political divides when people feel misunderstood.

Related Biases: False Consensus Effect and Curse of Knowledge.

Countermeasures: The antidote to the Illusion of Transparency is clear, explicit communication. Instead of assuming others can read your mind, verbalize your thoughts and emotions. Ask for feedback and clarifications.

Remember, the walls of your mental fortress are not made of glass.

Key Takeaways: The Illusion of Transparency might make us feel exposed and vulnerable, but most people are too absorbed in their own glass houses to notice. Learning to communicate clearly and effectively can minimize misunderstandings and improve social interactions.

References:

Gilovich, T., Savitsky, K., & Medvec, V. H. (1998). The illusion of transparency: Biased assessments of others' ability to read one's emotional states. Journal of Personality and Social Psychology, 75(2), 332-346.

Savitsky, K., & Gilovich, T. (2003). The illusion of transparency and the alleviation of speech anxiety. Journal of Experimental Social Psychology, 39(6), 618-625.

Illusion of Truth Effect

Definition: The Illusion of Truth Effect is a cognitive bias that suggests if we hear a statement often enough, we'll start to believe it's true, regardless of its actual veracity. Basically, it's the idea that familiarity breeds belief, not contempt.

Description: This bias exploits our tendency to mistake familiarity for truth. We are, for some unfathomable reason, wired to believe that if we've heard something repeatedly, it must have some merit. It's like getting a song stuck in your head—just because it's on repeat doesn't mean it's a good song.

Examples: In politics, a false statement repeated enough times might become accepted as truth. Similarly, in religion, a parable or belief, once continuously propagated, can become an incontestable truth. This effect also supports the rise of social myths, conspiracy theories, and misinformation spread on social platforms.

Impact: This bias can contribute to the spread and acceptance of misinformation and propaganda. It can impact social and political discourse, potentially influencing elections and policy. It can also lead to misinformed decisions and distorted perceptions of reality.

Related Biases: Availability Heuristic, Confirmation Bias, and Mere Exposure Effect.

Countermeasures: Cultivate a healthy skepticism and fact-check before accepting statements as truth. Diversity in information sources also helps. If you find yourself starting to believe that every cat secretly wants to kill its owner, you might want to diversify your reading materials.

Key Takeaways: The Illusion of Truth Effect reminds us that repetition does not equate to truth. To avoid falling for this bias, remember that variety is the spice of life—and a good way to keep cognitive biases in check.

References:

Begg, I., Anas, A., & Farinacci, S. (1992). Dissociation of processes in belief: Source recollection, statement familiarity, and the illusion of truth. Journal of Experimental Psychology: General, 121(4), 446.

Hasher, L., Goldstein, D., & Toppino, T. (1977). Frequency and the conference of referential validity. Journal of Verbal Learning and Verbal Behavior, 16(1), 107-112.

Illusory of Understanding

Definition: The Illusory of Understanding is a cognitive bias where people believe they understand a complex topic or process in greater detail than they actually do. Essentially, it's like claiming you can build a car when you only just mastered changing a tire.

Description: This charming little bias might convince you that understanding the gist of something means you've got the whole thing down pat. In reality, your knowledge could be as shallow as a kiddie pool, but hey, at least it's warm and comforting, right?

Examples: In the realm of politics, citizens often believe they understand complex policies or international relations when, in fact, their understanding is superficial. Similarly, in religious discussions, adherents might claim comprehensive understanding of their holy texts based on a few choice excerpts. Socially, it may be as simple as believing we understand the complexities of another person's experiences based on a single conversation or social media post.

Impact: This bias can lead to overconfidence and poor decision-making. It can lead to misunderstanding, miscommunication, and even conflict in political, religious, and social contexts.

Related Biases: Overconfidence Effect, Dunning-Kruger Effect, and Curse of Knowledge.

Countermeasures: To combat this bias, regularly seek out new learning opportunities, and approach complex topics with the understanding that they are, indeed, complex.

Practice humility, and remember: the more you know, the more you realize you don't know.

Key Takeaways: The Illusory of Understanding can lead us into a false sense of comprehension. It's always best to remember that any deep understanding is like a lasagna—layered, complex, and, hopefully, well-cooked.

References:

Rozenblit, L., & Keil, F. (2002). The misunderstood limits of folk science: an illusion of explanatory depth. Cognitive Science, 26(5), 521-562.

Fernbach, P. M., Rogers, T., Fox, C. R., & Sloman, S. A. (2013). Political extremism is supported by an illusion of understanding. Psychological Science, 24(6), 939-946.

Illusory Pattern Perception

Definition: Illusory Pattern Perception is the cognitive bias wherein individuals perceive meaningful patterns or connections in random or meaningless data. It's kind of like seeing constellations in the night sky. Sure, Orion's Belt is cool, but let's be honest, it's just a straight line.

Description: Our brain is a pattern-recognition machine. It's fantastic when distinguishing between a bear and a big rock, not so much when we start seeing secret codes in our alphabet soup. This bias taps into our innate desire for order and predictability. It's like the universe's version of a magic eye puzzle.

Examples: In the religious sphere, someone may interpret natural disasters as divine punishment for societal misdeeds. Politically, one might perceive patterns or conspiracies in independent events, like connecting the implementation of a new law to a minor stock market fluctuation. Socially, someone might interpret an acquaintance's sporadic responses to their messages as a sign of dislike, while there could be numerous other explanations.

Impact: Illusory Pattern Perception can lead to belief in conspiracy theories, superstitions, and pseudoscience, contributing to misinformed decisions and actions. On a broader scale, it can fuel societal divisions and misunderstandings.

Related Biases: Apophenia, Pareidolia, and Clustering Illusion.

Countermeasures: Cultivate critical thinking skills, question your assumptions, and always seek reliable sources of information. Remember, correlation does not equal causation, and sometimes, a cigar is just a cigar.

Key Takeaways: The universe might seem like it's full of hidden connections and patterns, but that doesn't mean it's trying to send you secret messages. Stay grounded in facts and reason, and don't let the stars dictate your life choices.

References:

Whitson, J. A., & Galinsky, A. D. (2008). Lacking control increases illusory pattern perception. Science, 322(5898), 115-117.

Foster, K. R., & Kokko, H. (2009). The evolution of superstitious and superstition-like behaviour. Proceedings of the Royal Society B: Biological Sciences, 276(1654), 31-37.

Impact Bias

Definition: Impact Bias is the tendency to overestimate the intensity and duration of emotional responses to future events. It's like assuming eating that entire tub of ice cream will permanently erase the pain of your breakup.

Description: This cognitive quirk comes from our tendency to underestimate our psychological resilience and forget the wide range of emotions we typically experience. Simply put, we're drama queens of our futures, predicting tragic operas instead of the sitcoms our lives usually are.

Examples: In religion, an adherent might overestimate the emotional regret they'd feel if they miss a religious ceremony. Politically, someone might anticipate being forever distraught if their favored candidate doesn't win an election. Socially, imagine you're expecting to be crushed by loneliness after a friend moves away, forgetting you have the ability (and Wi-Fi) to make new connections.

Impact: Impact Bias can lead to poor decision-making, as individuals might avoid beneficial actions due to overestimated negative emotional consequences, or pursue harmful ones because they overestimate the positive ones. Remember that ice cream scenario? Exactly.

Related Biases: Affective Forecasting, Overconfidence Bias.

Countermeasures: Practice mindfulness and emotional regulation. Understand that emotions are transitory and often less extreme than anticipated. Keep a diary of your emotional predictions and their real outcomes. Over time, this might convince you that you're no Nostradamus of emotions.

Key Takeaways: We tend to exaggerate our emotional responses to future events. Emotions are more like weather—changing often and rarely as extreme as predicted. So, brace for drizzles, not thunderstorms.

References:

Wilson, T. D., & Gilbert, D. T. (2003). Affective forecasting. Advances in Experimental Social Psychology, 35, 345-411.

Gilbert, D. T., Pinel, E. C., Wilson, T. D., Blumberg, S. J., & Wheatley, T. P. (1998). Immune neglect: A source of durability bias in affective forecasting. Journal of Personality and Social Psychology, 75(3), 617-638.

Impostor Syndrome

Definition: Impostor Syndrome refers to the psychological pattern in which individuals doubt their skills, talents, or accomplishments and have a persistent, internalized fear of being exposed as a "fraud". Much like an undercover agent worrying their cover might be blown at any moment, even if they're only infiltrating their own life.

Description: This isn't just a humblebrag in disguise. People with Impostor Syndrome genuinely believe they've only succeeded due to luck, timing, or deceiving others into thinking they're more competent than they actually are. It's as if they're forever preparing for an Oscar for "best person pretending to be themselves."

Examples: In religious contexts, a leader might worry they're not spiritual enough to lead, despite overwhelming evidence to the contrary. Politically, a newly-elected official may feel like a pretender in the midst of seasoned politicians. In social settings, someone may feel they don't "deserve" their friends, imagining they've somehow tricked these folks into liking them.

Impact: Impostor Syndrome can lead to stress, anxiety, low self-confidence, fear of failure, and even depression. It can hinder personal development, risk-taking, and contribute to burnout. Plus, it's a tremendous waste of worry about a reveal that's never going to come.

Related Biases: Self-Serving Bias, Dunning-Kruger Effect.

Countermeasures: Seek out supportive networks and mentorships, learn to internalize successes and externalize failures, and realize that perfection is about as realistic as a unicorn riding a dragon. Therapy or counseling can be helpful, too.

Key Takeaways: Impostor Syndrome is a pesky self-deprecating belief that discounts your achievements and skills. You're not an impostor in your own life, and believing you are is probably the only fraudulent thing about you.

References:

Clance, P. R., & Imes, S. A. (1978). The imposter phenomenon in high achieving women: Dynamics and therapeutic intervention. Psychotherapy: Theory, Research & Practice, 15(3), 241–247.

Sakulku, J., & Alexander, J. (2011). The impostor phenomenon. International Journal of Behavioral Science, 6(1), 73-92.

Inattentional Blindness

Definition: Inattentional blindness is a psychological lack of attention that isn't associated with vision defects or deficits. It's essentially the mental equivalent of "can't see the forest for the trees" — except sometimes, we can't even see the trees.

Description: This phenomenon occurs when an individual fails to perceive an unexpected stimulus in plain sight, simply because their attention is engaged with another task, event, or object. You might say, our brains are masterful at multitasking, except for when they're not. At all.

Examples: In politics, it could be ignoring a vital issue due to focus on a captivating scandal. Within religion, one might overlook the nuances of another's belief system, concentrating solely on their own doctrines. In social matters, we often miss the subtle signs of distress in others because we're too wrapped up in our own daily drama.

Impact: Inattentional blindness can limit our understanding, leading to poor decision-making and lack of empathy. It might cause you to miss the gorilla in the room, metaphorically speaking, and sometimes even literally (remember that selective attention test with the basketball passes?).

Related Biases: Change Blindness, Selective Perception, Confirmation Bias.

Countermeasures: Cultivating mindfulness and active observation can help. Try this: every now and then, take off the mental blinders and have a look around. You'll be amazed at what you've been missing.

Key Takeaways: Inattentional blindness makes us miss the obvious because our attention is elsewhere. The world is a smorgasbord of stimuli; don't fill up on breadsticks and miss the entrées.

References:

Mack, A., & Rock, I. (1998). Inattentional Blindness. MIT Press.

Simons, D. J., & Chabris, C. F. (1999). Gorillas in our midst: Sustained inattentional blindness for dynamic events. Perception, 28(9), 1059–1074.

Inductive Bias

Definition: Inductive bias is the tendency to generalize from specific instances to broader rules or patterns, essentially the mental equivalent of "once bitten, twice shy."

Description: Inductive reasoning allows us to make predictions about future events based on past experiences. However, this "predictive prowess" sometimes has us overstepping, leading us to make sweeping generalizations from a limited set of observations. It's like thinking all dogs are scary just because one chihuahua barked at you that one time.

Examples: Politically, it may manifest in judging a party's entire ethos based on one policy. In religion, a single doctrine or practice might color one's perception of an entire faith. Socially, the behavior of one individual might lead us to stereotype an entire group.

Impact: Inductive bias can lead to over-generalizations, stereotypes, and ultimately, inaccurate judgments. It's the cognitive equivalent of throwing the baby out with the bathwater, and who wants to do that?

Related Biases: Stereotyping, Hasty Generalization, Confirmation Bias.

Countermeasures: Embrace diversity and complexity. Remember that individual instances may not represent the whole. As Mark Twain might've said, "All generalizations are false, including this one."

Key Takeaways: Inductive bias is our tendency to generalize from specific instances, but life, like a good box of chocolates, is diverse and surprising. Don't judge a chocolate by its wrapper.

References:

Holland, J. H., Holyoak, K. J., Nisbett, R. E., & Thagard, P. R. (1989). Induction: Processes of Inference, Learning, and Discovery. MIT Press.

Wason, P. C. (1960). On the failure to eliminate hypotheses in a conceptual task. Quarterly Journal of Experimental Psychology, 12(3), 129–140.

Information Bias

Definition: Information bias is the tendency to seek more information, even when it will not affect action. Imagine buying a dozen books on weight loss, but never doing a single sit-up. Yeah, that.

Description: It's human to crave knowledge, and that's a good thing—until it's not. Information bias is the equivalent of 'paralysis by analysis'. We become so engrossed in acquiring more information that we forget to act or overvalue irrelevant data, making our decisions more complicated and less efficient.

Examples: Politically, one might delve into the voting history of a candidate from their first school election onwards. Religiously, one could research every doctrine of every denomination, rather than focusing on the core beliefs. In social issues, one might obsess over every news story on a topic, losing sight of the broader picture.

Impact: Information bias can lead to decision paralysis, wasted resources, and stress. Moreover, it can cause us to lose focus and become lost in an ocean of irrelevant data. Remember, not everything that counts can be counted, and not everything that can be counted counts.

Related Biases: Analysis Paralysis, Information Overload, Choice Overload.

Countermeasures: Identify the essential data needed for decision-making, and ignore the rest. Try to follow the principle of 'just-in-time' information – acquiring it when you need it, not hoarding it 'just-in-case'.

Key Takeaways: While information is power, too much can lead to power cuts. Learn to differentiate between what's necessary and what's superfluous. Remember, the tallest stacks of books are often in the homes of those who know the least.

References:

Scherer, L. D., & Lambert, A. J. (2009). Contrast effects in priming paradigms: Implications for theory and research on implicit attitudes. Journal of Personality and Social Psychology, 97(3), 383–403.

Von Neumann, J., & Morgenstern, O. (1944). Theory of Games and Economic Behavior. Princeton University Press.

Insensitivity to Sample Size

Definition: Insensitivity to sample size is the cognitive bias that occurs when people under-react to the size of a sample or ignore it altogether. If you think that polling five people is enough to predict an entire country's political leanings, you're likely a card-carrying member of this club.

Description: This bias stems from our misunderstanding of how statistics work. We often don't appreciate that larger samples are more representative of the whole population. Hence, we tend to overgeneralize based on small, potentially unrepresentative samples.

Examples: For instance, imagine a person asserting that a political candidate is guaranteed victory because all five of their friends are voting for them. Or someone abandoning their new diet because they didn't lose weight in the first two days. These are classic cases of our subject not appreciating the importance of sample size.

Impact: The impact of this bias is seen in poor decision-making, inaccurate predictions, and skewed perceptions in politics, religion, and social issues. A politician might pander to a vocal minority, mistaking it for the majority. Or a person might hastily label a new religious movement as a 'cult' based on a single negative report.

Related Biases: Neglect of Probability, Overgeneralization, Hasty Generalization.

Countermeasures: To mitigate this bias, we need to educate ourselves on the basics of statistical reasoning. Practice questioning the sample size and representativeness in the information you consume.

Key Takeaways: Bigger (sample sizes) are generally better. While small samples can give insights, they often fail to reflect the broader context. So next time you're ready to make a sweeping generalization, consider whether your sample is a slice or just a crumb.

References:

Tversky, A., & Kahneman, D. (1971). Belief in the law of small numbers. Psychological Bulletin, 76(2), 105–110.

Kahneman, D., & Tversky, A. (1992). Advances in prospect theory: Cumulative representation of uncertainty. Journal of Risk and Uncertainty, 5(4), 297–323.

Introspection Illusion

Definition: The Introspection Illusion is a cognitive bias where individuals overestimate their ability to accurately understand their own mental processes. It's like thinking you're a walking, talking Freud, all capable of decoding your subconscious, when, in reality, you're more of a bumbling Watson, trying to keep up with Sherlock's leaps of deduction.

Description: The heart of the introspection illusion lies in our confidence in our self-analysis. We assume we can peer into the murky depths of our minds and dredge up the 'why' behind our actions, beliefs, and emotions. But often, our introspection can be as clear as mud, leading us to make inaccurate assumptions about ourselves.

Examples: A politician might believe they're unbiased, convinced their policy choices stem solely from clear logic and not party lines. Or a social media user might consider themselves impervious to viral trends, only to find they've inexplicably taken up flossing (the dance, not the dental hygiene routine).

Impact: This bias affects our self-awareness, personal development, and relationships. In politics, it can create blind spots in politicians, preventing them from acknowledging their biases. In social issues, it can hinder open dialogue and change, with individuals unwilling to accept the unconscious biases they might harbor.

Related Biases: Self-Serving Bias, Illusion of Transparency, Blind Spot Bias.

Countermeasures: To counteract this bias, it's helpful to seek feedback from others and to be open to criticism. Also, practicing mindfulness can increase self-awareness and improve introspective accuracy.

Key Takeaways: Even though we live with ourselves 24/7, our understanding of our mental processes can often be as sketchy as a flipbook drawn in invisible ink. So, let's be humble mental detectives, always open to the possibility that our first-person perspective might not tell the full story.

References:

Pronin, E., Lin, D. Y., & Ross, L. (2002). The bias blind spot: Perceptions of bias in self versus others. Personality and Social Psychology Bulletin, 28(3), 369–381.

Wilson, T. D., & Dunn, E. W. (2004). Self-knowledge: Its limits, value, and potential for improvement. Annual Review of Psychology, 55, 493–518.

In-group Bias

Definition: In-group Bias, often deemed the ultimate "us versus them" mindset, is the tendency to favor our own group members over those from other groups. It's like believing your family makes the best Thanksgiving turkey just because, well, they're your family.

Description: This cognitive bias arises from our proclivity to categorize people into groups. It's the psychological version of the neighborhood block party, but with less barbecue and more favoritism. We're more likely to trust, cooperate with, and attribute positive qualities to those in our in-group, even if the only thing we have in common is our love for cat memes.

Examples: For a political example, consider party members overlooking flaws in their own candidate while scrutinizing the opposing party's candidate for similar issues. On a broader social scale, sports fans often view their team's rivals with irrational disdain, despite the only real difference being the color of their jerseys.

Impact: In-group bias can fuel harmful stereotypes and discrimination, exacerbating social divides. It can result in partisan politics, where policies are passed not based on merit but party affiliation. It's a recipe for division, substituting fairness and unity with bias and blind loyalty.

Related Biases: Out-group Homogeneity Bias, Confirmation Bias, Bias Blind Spot.

Countermeasures: Cultivating empathy and encouraging diversity can mitigate this bias. Additionally, focusing on shared goals or commonalities can help bridge the gap between 'us' and 'them'.

Key Takeaways: While it's human nature to form groups, it's vital to remember that this doesn't necessitate an "us vs. them" scenario. Because whether it's politics, sports, or the last slice of pizza, favoring our group blindly can often lead to unfairness and unnecessary division.

References:

Brewer, M. B. (1999). The psychology of prejudice: Ingroup love or outgroup hate? Journal of Social Issues, 55(3), 429-444.

Tajfel, H., & Turner, J. C. (1986). The social identity theory of intergroup behavior. In S. Worchel & W. G. Austin (Eds.), Psychology of intergroup relations (2nd ed., pp. 7-24). Nelson-Hall.

Irrelevant Alternatives Fallacy

Definition: The Irrelevant Alternatives Fallacy, also known as the Decoy Effect or Attraction Effect, is a cognitive bias in which an individual's preference between two options changes in the presence of a third, seemingly irrelevant option.

Description: Underlying this quirky fallacy is the way our brains handle choices. The unassuming gray matter atop your neck enjoys comparison, you see. It revels in context. Give it two options, it will compare; but throw in an irrelevant third, and suddenly, it seems the original duo have been on America's Got Talent, and the newcomer is the buzzer. This bias can make us choose an option we wouldn't have considered if it weren't for the irrelevant alternative.

Examples: Picture a local election. Two candidates, A and B, have your interest. Candidate A is a fiscal conservative with a penchant for environmental policy, while candidate B leans liberal and emphasizes education. Now, enter candidate C, an under-qualified, obscure candidate with views so far left they've almost looped around. Suddenly, candidate B seems like the more centrist, reasonable choice, even if your preferences between A and B haven't changed one bit.

Impact: The Irrelevant Alternatives Fallacy can seriously distort decision-making, leading people to make choices they wouldn't otherwise. In politics, it can manipulate electoral results. In consumer behavior, it can nudge you towards buying that more expensive, but "worth it", deal.

Related Biases: This fallacy is related to the Anchoring Effect, where people rely too heavily on the first piece of information they encounter, and the Framing Effect, where people decide based on the way information is presented.

Countermeasures: Recognize this bias and critically analyze your choices. Strip away the fluff. Consider the value of each choice individually, and remember, if an option wasn't valuable to begin with, its presence shouldn't shift your preferences.

Key Takeaways: The Irrelevant Alternatives Fallacy can make you dance to the tune of irrelevant options. Always remember, when making decisions, to base your choices on their intrinsic value and not their relative attractiveness.

References:

Huber, J., Payne, J. W., & Puto, C. (1982). Adding Asymmetrically Dominated Alternatives: Violations of Regularity and the Similarity Hypothesis. Journal of Consumer Research, 9(1), 90–98. https://doi.org/10.1086/208899

Pettibone, J. C., & Wedell, D. H. (2007). Testing Alternative Explanations of Phantom Decoys: An Analysis of Decision Weights. Journal of Behavioral Decision Making, 20(3), 323–341. https://doi.org/10.1002/bdm.567

Irrational Escalation

Definition: Irrational Escalation, or Escalation of Commitment, is a cognitive bias that causes individuals or groups to continue on a failing course of action, simply because they have already invested significant resources (time, money, effort) into it.

Description: Essentially, this bias is the manifestation of that classic saying, "throwing good money after bad." Why would we do such a thing, you might ask? Well, it's mostly because we humans have an aversion to loss, a firm belief in sunk cost, and an ego larger than the collective Twitter universe. These ingredients mix up a pretty potent cocktail of irrational escalation.

Examples: The political realm is replete with examples. Picture a nation, having declared war for a cause that's lost its relevance or support, yet continuing the conflict because of the resources already invested. It's a bit like lighting your money on fire because it got slightly singed.

Impact: Irrational escalation can lead to disastrous consequences, such as financial ruin, persistent conflict, or wasted resources. In social issues, it can perpetuate harmful norms or behaviors.

Related Biases: This bias is strongly linked with the Sunk Cost Fallacy, where past investments impact future decisions, and Loss Aversion, where the potential for loss is weighted more heavily than an equivalent gain.

Countermeasures: To mitigate this bias, reevaluate your course regularly. Be ready to accept losses, because hey, everyone trips now and then. Seeking external, objective opinions can also help keep you grounded in reality, instead of being swayed by the glamour of past investments.

Key Takeaways: Be mindful of irrational escalation. Don't let past investments dictate your future decisions. Sometimes, it's wise to cut your losses and move on. Remember, there's no point pouring good coffee into a leaky mug.

References:

Staw, B. M. (1976). Knee-deep in the big muddy: A study of escalating commitment to a chosen course of action. Organizational behavior and human performance, 16(1), 27-44. https://doi.org/10.1016/0030-5073(76)90005-2

Whyte, G. (1986). Escalating commitment to a course of action: A reinterpretation. Academy of management review, 11(2), 311-321. https://doi.org/10.5465/amr.1986.4282625

Just-Noticeable Difference (JND)

Definition: The Just-Noticeable Difference, also known as Weber's Law, is a concept in psychology that explains the minimum amount a stimulus intensity must be changed in order to produce a noticeable variation in sensory experiences.

Description: Let's say you're carrying a bag of groceries. It's heavy, but manageable. Then someone adds a grape to it. Do you notice? Probably not. But if someone adds a watermelon, you're likely to feel the difference. JND is that threshold where a grape becomes a watermelon. Metaphorically speaking, of course.

Examples: Politically, let's say a government subtly curtails civil liberties. If changes are small and gradual, citizens may not notice until the landscape of freedom is drastically altered. Think of the proverbial frog in the slowly heated pot of water. Spoiler: it doesn't end well for the frog.

Impact: The JND has significant implications for perception and decision-making. It can lead to a gradual acceptance of changes that, if implemented abruptly, would cause uproar or resistance. This can occur in any domain, including religion, politics, or social norms.

Related Biases: JND is related to Change Blindness, where observers fail to notice changes in their visual field, and the Boiling Frog Phenomenon, where gradual changes go unnoticed until it's too late.

Countermeasures: Awareness is key. Keeping an informed perspective about ongoing changes, no matter how small, helps in recognizing the potential long-term impact. It's like keeping an eye on the grocery bag so no one sneaks a watermelon into it.

Key Takeaways: Keep tabs on the small changes, because they often add up. As with carrying groceries or boiling frogs, noticing subtle shifts early on can prevent an unexpected heavy load or a hot bath later.

References:

Goldstein, E. B. (2019). Sensation and perception. Boston, MA: Cengage Learning.

Weber, E. H. (1834). De pulsu, resorptione, auditu et tactu: Annotationes anatomicae et physiologicae. Koehler.

Just-World Hypothesis

Definition: The Just-World Hypothesis is a cognitive bias that leads people to believe that the world is inherently fair, causing them to rationalize an otherwise inexplicable injustice as deserved by the recipient.

Description: Imagine a world where all the good guys get rich, find love, and live to a ripe old age, while the villains end up poor, lonely, and meet untimely ends. Sounds like a typical Hollywood movie, right? That's the Just-World Hypothesis for you. It's comforting to believe that justice always prevails, but let's be real, life's not a movie and the script often doesn't follow the fairy-tale narrative.

Examples: If a politician you dislike is involved in a scandal, the Just-World Hypothesis might lead you to say, "Well, that's karma for you!" But if a politician you support is in the same boat, you might blame it on a conspiracy or bad luck. In both scenarios, the bias helps maintain your belief in a just world where people get what they deserve.

Impact: This bias can lead to victim-blaming and can stifle the impulse to challenge injustice. It's easier to believe that a poor person must have done something to deserve their plight than to accept the uncomfortable truth of systemic inequality.

Related Biases: The Fundamental Attribution Error (attributing others' actions more to their character than to their circumstances) can be seen as a sidekick to the Just-World Hypothesis.

Countermeasures: Question your assumptions. Is someone really to blame for their misfortune, or are there systemic factors at play? Embrace the reality of randomness and unfairness in life. It's okay, life's still worth living. I promise.

Key Takeaways: The world isn't a vending machine where you insert virtue and receive a soda of success. Injustice exists, and acknowledging it is the first step toward fighting it.

References:

Lerner, M. J., & Miller, D. T. (1978). Just world research and the attribution process: Looking back and ahead. Psychological Bulletin, 85(5), 1030–1051.

Hafer, C. L., & Bègue, L. (2005). Experimental Research on Just-World Theory: Problems, Developments, and Future Challenges. Psychological Bulletin, 131(1), 128-167.

Law of the Instrument

Definition: The Law of the Instrument, also known as Maslow's Hammer, is a cognitive bias that leads to over-reliance on a familiar tool or perspective, encapsulated in the phrase, "If all you have is a hammer, everything looks like a nail."

Description: It's like being stuck in a culinary rut. You've just learned how to make a mean lasagna. So, you make it Monday. Then Tuesday. By Wednesday, even the dog has lost interest. By Sunday, you're seriously considering opening a lasagna-themed restaurant. This is the Law of the Instrument, where your newfound talent, your 'hammer,' turns every mealtime into 'nail' time. This bias leads us to overuse a particular approach simply because it's familiar or handy.

Examples: In politics, if a leader's main strategy is to appeal to patriotism, they might try to frame every issue, from economics to health, as a matter of national pride. It might work on Independence Day, but it's going to be a tough sell during tax season.

Impact: Over-reliance on a single tool or perspective can limit creativity, inhibit problem-solving, and lead to one-dimensional strategies. It can also cause us to overlook more effective solutions, simply because they're not within our immediate toolkit.

Related Biases: Functional fixedness, a type of cognitive bias that limits a person to using an object only in the way it is traditionally used, is a close cousin of the Law of the Instrument.

Countermeasures: Cultivate a diverse set of tools, perspectives, and strategies. Challenge yourself to think outside the box. And remember, while lasagna is great, so is a good taco, salad, or sushi roll.

Key Takeaways: Diversity in your mental toolbox allows for versatile and innovative problem-solving. Relying on the same old hammer not only makes you a boring dinner host but also limits your potential for truly nailing solutions.

References:

Maslow, A. H. (1966). The Psychology of Science. A Reconnaissance. Chapel Hill: Maurice Bassett Publishing. pp. 15–16.

Kaplan, Abraham (1964). "The Conduct of Inquiry: Methodology for Behavioral Science". Scranton, Pennsylvania: Chandler Publishing Co. pp. 28-29.

Law of Triviality

Definition: The Law of Triviality, also known as "Bikeshedding," is a cognitive bias that causes people to spend a disproportionate amount of time on trivial, easy-to-grasp issues, often at the expense of more critical, complex ones.

Description: Picture this: your local town council has two items on the agenda. One: choosing the color of a new bike shed. Two: planning the town's nuclear waste management strategy. Guess which one sparks a lively debate? Yep, the bike shed. Why? Because choosing a color is easy and everyone has an opinion. Radioactive waste? Not so much. That's the Law of Triviality in action: trivial matters are often given more attention than complex ones because they're simply easier to understand and discuss.

Examples: In politics, a debate over the menu at the White House State Dinner might garner more public interest than the intricacies of healthcare reform. In religious circles, a debate over the proper attire for Sunday service could sidetrack deeper theological discussions.

Impact: While it's fun to argue about bike sheds (or White House dinners), focusing on trivialities can lead to neglect of more important matters, hindering progress and decision-making on critical issues.

Related Biases: The Law of Triviality is related to Parkinson's Law of Triviality, and it's a specific form of the Availability Heuristic, where we focus on the information most readily available to us.

Countermeasures: Prioritize issues based on their importance, not ease of understanding. Allocate time proportionally to the gravity of the issues at hand. And remember, the color of the bike shed won't matter if the town is glowing green.

Key Takeaways: Keep your eye on the ball, or in this case, the nuclear waste. Don't allow easy but inconsequential issues to distract from the truly significant ones.

References:

Parkinson, C. Northcote (1957). Parkinson's Law: The Pursuit of Progress. John Murray. p. 4.

Kahneman, D., & Tversky, A. (1973). On the psychology of prediction. Psychological Review, 80(4), 237–251.

Less-is-Better Effect

Definition: The Less-is-Better Effect is a cognitive bias where, in certain circumstances, less is perceived as more. Essentially, it's the opposite of an all-you-can-eat buffet philosophy.

Description: This psychological phenomena can make a four-piece silver dinner set seem more valuable than a 24-piece set with a few tarnished spoons. The completeness and perfection of the smaller set outweigh the greater quantity of the larger set. It's as if our brains pull out a mental magnifying glass and zoom in on the imperfections when there's more to scrutinize.

Examples: In the political realm, a candidate who makes a few focused, concise promises may be seen as more trustworthy than one who promises the moon and stars (and possibly a free unicorn). In religious contexts, a minimalist lifestyle, such as that practiced by monks or nuns, may be perceived as more virtuous than a more extravagant existence.

Impact: The Less-is-Better Effect can lead to paradoxical decision-making, as it challenges the conventional notion that more is always better. This can have significant implications in various aspects of life, from consumer choices to policy decisions.

Related Biases: This bias is closely related to the Paradox of Choice, where too many options can lead to decision paralysis or dissatisfaction, and the Scarcity Principle, which suggests that limited availability can make things seem more valuable.

Countermeasures: Be mindful of the quality-quantity balance and don't let the illusion of less-is-better cloud your judgment. And remember: the number of unicorns a politician promises is less important than their capacity to deliver on their commitments.

Key Takeaways: Less can sometimes be more, but it's crucial to evaluate the intrinsic value of what's on offer, rather than being swayed by the deceptive charm of scarcity.

References:

Hsee, C. K. (1998). Less is better: When low-value options are valued more highly than high-value options. Journal of Behavioral Decision Making, 11(2), 107–121.

Schwartz, B. (2004). The paradox of choice: Why more is less. Harper Perennial.

Cialdini, R. B. (2009). Influence: Science and practice. Pearson education.

Likability Bias

Definition: Likability Bias is the cognitive tendency to hold positive views and give preferential treatment to people we like, irrespective of their objective merits or demerits.

Description: It's not rocket science, really. We tend to be biased towards people we like. This affectionate lens through which we view our favorites can skew our objectivity, much like wearing rose-colored glasses at a tomato festival. Everything just seems more...well, rosy.

Examples: A charismatic political leader who tickles our funny bone might get our vote over a less charming, yet potentially more qualified, candidate. In religious contexts, people often favor interpretations of scripture that are presented by more likable religious leaders. On social issues, popular celebrities often wield considerable influence over public opinion, whether or not they're actually knowledgeable on the topic.

Impact: Likability Bias can lead to poor decision-making, as we let our personal feelings override logical and rational analysis. This can perpetuate inequality and prejudice, as our unconscious preferences for likable individuals can often be influenced by societal stereotypes and norms.

Related Biases: Similar biases include the Halo Effect, where we perceive someone as good in all areas because they excel in one, and Confirmation Bias, where we favor information that confirms our preexisting beliefs.

Countermeasures: A healthy dose of self-awareness and critical thinking can go a long way in countering Likability Bias. Consider people's qualifications and positions objectively, and try to separate the message from the messenger.

Key Takeaways: Likability can greatly influence our perceptions and decisions. It's essential to separate the person from the task at hand and avoid letting our fondness for someone cloud our judgement.

References:

Cialdini, R. B. (2009). Influence: Science and practice. Pearson education.

Nickerson, R. S. (1998). Confirmation bias: A ubiquitous phenomenon in many guises. Review of General Psychology, 2(2), 175–220.

Nisbett, R. E., & Wilson, T. D. (1977). The halo effect: Evidence for unconscious alteration of judgments. Journal of Personality and Social Psychology, 35(4), 250–256.

Licensing Effect

Definition: The Licensing Effect, or moral self-licensing, is the cognitive bias that allows individuals to justify morally dubious decisions or actions after having previously acted in positive or ethical ways.

Description: So, you've been a good Samaritan, helping an elderly person cross the road, recycling your waste, and feeding stray cats. Well done, you! But hold on, don't start planning that bank heist just yet. Unfortunately, our minds have a funny way of using good deeds as a license to commit not-so-good deeds. This phenomenon, known as the Licensing Effect, is like getting a 'get out of jail free card' from Monopoly for real life. Only, it's all in your head, and the real world doesn't work that way.

Examples: A politician fighting for climate change may feel justified in flying private jets, citing their 'green credentials.' Religious leaders might turn a blind eye to their own moral transgressions, pointing to their service to their community. A person who has been dieting may reward themselves with a high-calorie treat, negating their prior efforts.

Impact: The Licensing Effect can hinder societal progress by allowing individuals and institutions to justify unethical behaviors, thereby maintaining status quo in politics, religion, and social issues. It can also harm personal development by sabotaging our own best efforts to improve.

Related Biases: The Licensing Effect is related to Cognitive Dissonance, where we experience discomfort due to holding contradictory beliefs, and Confirmation Bias, which leads us to favor information that confirms our pre-existing beliefs.

Countermeasures: Being aware of this bias and taking steps to counteract it can help. Instead

of seeing good behavior as a pass for less desirable actions, strive for consistency in your actions. Employ critical thinking and strive to make decisions based on their own merits rather than being influenced by unrelated actions.

Key Takeaways: The Licensing Effect can lead us to excuse bad behavior due to past good actions. It's essential to strive for consistent ethical behavior and avoid using past good deeds as a license to lapse in the present.

References:

Khan, U., & Dhar, R. (2006). Licensing Effect in Consumer Choice. Journal of Marketing Research, 43(2), 259–266.

Merritt, A. C., Effron, D. A., & Monin, B. (2010). Moral Self-Licensing: When Being Good Frees Us to Be Bad. Social and Personality Psychology Compass, 4(5), 344–357.

Festinger, L. (1962). Cognitive dissonance. Scientific American, 207(4), 93-107.

Loss Aversion

Definition: Loss aversion is the cognitive bias where individuals tend to prefer avoiding losses over acquiring equivalent gains. It's like the old saying goes, "a bird in the hand is worth two in the bush," or was it "losing a pound hurts twice as much as gaining one feels good?" Either way, you get the picture.

Description: Loss aversion makes our brain go into overdrive when we sense the risk of loss. This probably made a lot of sense when losing a piece of fruit could mean starving, but these days, it can just as easily lead us astray. In fact, it's like we have an internal accountant that inexplicably values losses more than gains, making our financial decision-making equivalent to a toddler's trade negotiations.

Examples: In politics, a candidate may choose to stick to an unpopular policy because they fear the loss of face in changing their stance more than the potential gain in popularity. In religion, an individual may adhere to certain restrictive practices, not out of genuine belief, but because they fear the loss of community support. Socially, we might stick to our routine even when exciting opportunities arise, simply because we fear losing our comfortable status quo.

Impact: Loss aversion can seriously skew decision-making processes in any sphere, causing us to miss out on potentially beneficial opportunities due to the overvaluation of potential loss. It can lead to resistance to change and innovation, stalling progress in critical areas like politics, religion, and social issues.

Related Biases: Loss aversion is closely related to the Endowment Effect (overvaluing things we own) and the Sunk Cost Fallacy (continuing a behavior due to previously invested resources).

Countermeasures: Recognizing loss aversion can help reduce its impact. Consider potential gains and losses in a balanced way, and don't let the fear of loss blind you to potential benefits.

Key Takeaways: Loss aversion can lead us to make irrational decisions based on fear of loss rather than an objective evaluation of potential gains and losses. Aim for balance and don't let the fear of losing out hold you back.

References:

Kahneman, D., Knetsch, J. L., & Thaler, R. H. (1991). Anomalies: The endowment effect, loss aversion, and status quo bias. The Journal of Economic Perspectives, 5(1), 193-206.

Tversky, A., & Kahneman, D. (1991). Loss aversion in riskless choice: A reference-dependent model. The Quarterly Journal of Economics, 106(4), 1039-1061.

Arkes, H. R., & Blumer, C. (1985). The psychology of sunk cost. Organizational Behavior and Human Decision Processes, 35(1), 124-140.

Mere Exposure Effect

Definition: The mere exposure effect is the cognitive bias that propels people to develop a preference for things merely because they are familiar with them. It's like hearing the same song on repeat until you start humming it yourself, even if it's not your musical taste. So much for personal preference!

Description: This effect suggests that repeated exposure to a stimulus increases our liking for it. So, if you're ever wondering why that mediocre movie feels better on the third watch, you have this cognitive bias to blame. Essentially, we're cognitive creatures of habit, so we tend to like things that we don't have to strain our brains over—familiar equals less work for the grey matter, apparently.

Examples: In politics, a candidate with greater media coverage can garner more votes simply due to increased familiarity. In religion, repeated exposure to certain practices or beliefs may deepen faith or commitment. In social issues, the public's acceptance of previously stigmatized or unfamiliar concepts can grow over time with increased exposure.

Impact: The mere exposure effect can impact our decisions in ways we're not consciously aware of, biasing us towards the familiar and away from the unknown. This can limit our openness to new experiences, ideas, or perspectives, reinforcing societal norms and status quos.

Related Biases: This effect is related to the familiarity heuristic (where familiar items are seen as superior to those that are unfamiliar) and the availability heuristic (relying on immediate examples that come to mind).

Countermeasures: Developing awareness of the mere exposure effect can help mitigate its influence. Try to stay open to new experiences, ideas, and perspectives, and don't shy away from the unfamiliar simply because it's... well, unfamiliar.

Key Takeaways: The mere exposure effect can cause us to favor the familiar and shun the unfamiliar, limiting our openness to new experiences. So, stay curious and don't let the comfort of familiarity become a mental straitjacket.

References:

Zajonc, R. B. (1968). Attitudinal effects of mere exposure. Journal of Personality and Social Psychology, 9(2), 1-27.

Bornstein, R. F. (1989). Exposure and affect: Overview and meta-analysis of research, 1968-1987. Psychological Bulletin, 106(2), 265-289.

Monahan, J. L., Murphy, S. T., & Zajonc, R. B. (2000). Subliminal mere exposure: Specific, general, and diffuse effects. Psychological Science, 11(6), 462-466.

Memory Bias

Definition: Memory bias is a cognitive bias that skews or distorts our recollection of past events, often in ways that serve our current beliefs, attitudes, or self-image.

Description: The human memory is no flawless recording device, contrary to what some courtroom dramas would have you believe. In reality, it's more like an eccentric artist - prone to exaggeration, fond of dramatic flair, and definitely not a fan of the mundane or contradictory. Memory bias occurs when our recollections are unconsciously influenced or colored by our present circumstances, beliefs, and emotions. These retrospective distortions can take various forms, from recalling events as more extreme or important than they were, to conveniently forgetting information that doesn't match our current worldview.

Examples: You might be wondering where this bias crops up. Well, fancy a stroll through the gardens of religion, politics, and social issues?

In the realm of religion, memory bias might take the form of a faithful person recalling past prayers as being invariably followed by positive outcomes. In politics, supporters of a particular party might remember their candidate's successes in Technicolor detail while their missteps seem to be lost in a mysterious, conveniently placed fog. Social issues are not immune either - folks might remember protests or movements they support as peaceful and impactful, whereas opposing ones were, surely, nothing but chaotic rabble-rousing.

Impact: The effects of memory bias can be as subtle as a tweak in a bedtime story or as seismic as an altered political landscape. It impacts our beliefs, decisions, and interpretations of reality, often reinforcing existing prejudices and inhibiting objective understanding. In effect, it can create a self-sustaining cycle: we shape our memories based on our beliefs, then use those biased memories to justify and strengthen the very same beliefs.

Related Biases: Memory bias is like the popular kid at the Cognitive Biases High - it's connected to a host of others. These include confirmation bias (favoring information that aligns with our beliefs), self-serving bias (perceiving oneself in an overly favorable manner), and hindsight bias (aka the "I-knew-it-all-along" effect).

Countermeasures: Minimizing memory bias is like trying to herd cats, a challenge to be sure, but not impossible. Encourage skepticism and critical thinking. Keep a record of important events to reference later. Practice mindfulness and self-awareness about the subjective nature of your memories. And if all else fails, remember that your memory isn't a documentary, but more of a Hollywood blockbuster - it may be entertaining, but it doesn't necessarily stick to the facts.

Key Takeaways: Your memory can be a tricky character, often misrepresenting the past in favor of your present self. Be aware of its penchant for bias, particularly in matters of religion, politics, and social issues, and always strive for critical thinking and evidence-based decision making.

References:

Schacter, D. L., & Addis, D. R. (2007). The cognitive neuroscience of constructive memory: Remembering the past and imagining the future. *Philosophical Transactions of the Royal Society B: Biological Sciences, 362*(1481), 773-786. doi:10.1098/rstb.2007.2087

Loftus, E. F. (2005). Planting misinformation in the human mind: A 30-year investigation of the malleability of memory. *Learning & Memory, 12*(4), 361-366. doi:10.1101/lm.94705

Roese, N. J., & Vohs, K. D. (2012). Hindsight Bias. *Perspectives on Psychological Science, 7*(5), 411–426. https://doi.org/10.1177/1745691612454303

Misattribution of Arousal

Definition: Misattribution of arousal is a psychological phenomenon where people misidentify the source of their emotional states, ascribing their feelings to an incorrect cause.

Description: In life, things get heated. Quite literally for spicy food aficionados. But when it comes to emotions, things get spicy in a different way. Our brains, the great connectors of dots, sometimes draw lines from A to Z when we're feeling emotionally charged, ignoring all the letters in between. This is the misattribution of arousal - when our high emotional states are incorrectly linked to a cause that seems plausible, but isn't the real catalyst.

Examples: Let's weave through the vibrant tapestry of religion, politics, and social issues for examples.

In religion, one might attend a high-energy service, complete with powerful music and dynamic oration. They leave feeling profoundly moved and ascribe this to a divine encounter, when it might simply be a reaction to the emotional atmosphere.

In politics, a passionate speech or heated debate can arouse strong feelings. This emotional charge is often attributed to the speaker's rhetoric, leading to inflated perceptions of their charisma or competency.

And in social issues, a particularly emotional documentary about climate change may spur feelings of guilt or concern, which are attributed to personal responsibility rather than the cinematic skills of the filmmakers.

Impact: Misattribution of arousal can lead to warped perceptions and flawed decisions. It can fuel extremism in religion or politics as emotional arousal is mistaken for divine inspiration or ideological conviction. In social issues, it can create unnecessary guilt or anxiety, or on the flip side, a false sense of heroism or virtue.

Related Biases: The misattribution of arousal rubs shoulders with other biases such as the Fundamental Attribution Error (overemphasizing personal characteristics and ignoring situational factors) and the Halo Effect (the tendency to allow one trait, such as attractiveness, to overshadow other traits).

Countermeasures: The fight against misattribution of arousal isn't a lost cause - it just requires a bit of emotional detective work. Practice mindfulness to tune into your feelings, and ask yourself, "Why am I really feeling this way?" You'll often find there's more to the story than meets the eye... or the heart.

Key Takeaways: Our brains are storytellers, and they often like to spin tales about why we feel the way we do. Be aware of this tendency to misattribute arousal, particularly in charged situations like religious events, political debates, or social issue discussions. Your emotions may be real, but their supposed cause might just be a convincing work of fiction.

References:

Zillmann, D. (1971). Excitation transfer in communication-mediated aggressive behavior. *Journal of Experimental Social Psychology, 7*(4), 419-434. doi:10.1016/0022-1031(71)90075-8

Dutton, D. G., & Aron, A. P. (1974). Some evidence for heightened sexual attraction under conditions of high anxiety. *Journal of Personality and Social Psychology, 30*(4), 510–517. https://doi.org/10.1037/h0037031

Mandler, G., Mandler, J. M., & Uviller, E. T. (1958). Autonomic feedback: The perception of autonomic activity. *The Journal of Abnormal and Social Psychology, 56*(3), 367–373. https://doi.org/10.1037/h0040660

Misinformation Effect

Definition: The misinformation effect refers to the phenomenon where a person's recall of episodic memories becomes less accurate due to post-event information, in simpler terms, people misremember things due to misleading information presented after the fact.

Description: Our memories, much like a competitive game of telephone, are subject to distortions and amendments. The misinformation effect illustrates this rather unsettling truth. It happens when our recall of past events is influenced by misleading information presented after the original event - a sort of post-production editing, if you will, except there's no Oscar for this kind of editing.

Examples: Now, let's explore the rabbit holes of religion, politics, and social issues to highlight this effect.

Within the context of religion, the misinformation effect might take hold when a religious leader introduces new interpretations or doctrines that alter a follower's memory of their original religious texts or teachings.

In politics, imagine a politician caught in a scandal. The initial report is quite damning, but subsequent spin by the politician's team introduces new details that muddy the waters. Over time, people's recollections of the original scandal might be swayed by this misleading information.

On the social issues front, consider how the narrative of historical events can change over time. For instance, with retellings and reinterpretations, peaceful protests may be remembered as violent riots, or vice versa.

Impact: The misinformation effect can shape our understanding of the world, often not for the better. It can rewrite history, distort truth, and perpetuate false narratives in religion, politics, and social issues. It shapes not just individual perceptions, but also collective memory, which can lead to societal misunderstanding and conflict.

Related Biases: This effect holds hands with other biases like the confirmation bias (prioritizing information that confirms our beliefs) and memory bias (altering memories based on current beliefs).

Countermeasures: Battling the misinformation effect requires a good dose of skepticism and a commitment to truth-seeking. Always verify new information, especially if it alters your understanding of past events. Encourage multiple sources, and remember, just because a detail is new doesn't mean it's true.

Key Takeaways: Our memories, charmingly fickle, can be hoodwinked by post-event misinformation, changing our perception of past events. It's essential to keep a discerning eye, especially in matters of religion, politics, and social issues. As they say, those who cannot remember the past are condemned to... well, misremember it.

References:

Loftus, E. F. (2005). Planting misinformation in the human mind: A 30-year investigation of the malleability of memory. *Learning & Memory, 12*(4), 361-366. doi:10.1101/lm.94705

Frenda, S. J., Nichols, R. M., & Loftus, E. F. (2011). Current Issues and Advances in Misinformation Research. *Current Directions in Psychological Science, 20*(1), 20–23. https://doi.org/10.1177/0963721410396620

Zaragoza, M. S., & Mitchell, K. J. (1996). Repeated exposure to suggestion and the creation of false memories. *Psychological Science, 7*(5), 294-300. doi:10.1111/j.1467-9280.1996.tb00377.x

Money Illusion

Definition: The money illusion is a cognitive bias where people think of currency in nominal terms, rather than real terms. This leads to the incorrect perception that nominal changes in price, wages, or market value equate to real changes, without accounting for inflation or deflation.

Description: Like a magician's sleight of hand, the money illusion makes us believe that what we see is what we get. But when it comes to money, face value doesn't always reflect actual value. This bias makes us see the world in nominal dollars and cents, ignoring pesky details like inflation or deflation that can alter purchasing power.

Examples: In our colorful mosaic of religion, politics, and social issues, the money illusion works its magic.

In religion, consider a church that sees an increase in tithing and assumes its congregation's devotion has risen, ignoring the inflationary environment which simply means people are giving the same percentage from bigger nominal incomes.

In politics, politicians might celebrate a nominal increase in minimum wage as a win for low-income workers. However, if inflation has risen at a higher rate, the purchasing power of that wage might have actually decreased.

When it comes to social issues, consider a charity that receives more donations year over year and celebrates its growing impact, without considering inflation. The actual purchasing power – and thus the real impact – may have stayed the same or even diminished.

Impact: The money illusion can skew our understanding of financial well-being and prosperity. It can lead to flawed policies, misguided donations, and a false sense of financial security or generosity. The cost of living can quietly rise while our perceived wealth remains the same, leading to an invisible erosion of purchasing power.

Related Biases: The money illusion is closely related to the anchoring bias (relying too heavily on initial information) and the framing effect (being influenced by the way information is presented).

Countermeasures: Beating the money illusion requires donning a pair of economic x-ray specs. Always consider the context of inflation or deflation when interpreting monetary values. Consult inflation calculators or financial advisors to help adjust your perception to reality.

Key Takeaways: Don't be fooled by the nominal face value of money. Inflation and deflation can make a big difference in the real value of money over time. It's not just about how much you have, but what you can do with it.

References:

Shafir, E., Diamond, P., & Tversky, A. (1997). Money Illusion. *The Quarterly Journal of Economics, 112*(2), 341-374. https://doi.org/10.1162/003355397555208

Fehr, E., & Tyran, J. R. (2001). Does Money Illusion Matter? *The American Economic Review, 91*(5), 1239-1262. https://doi.org/10.1257/aer.91.5.1239

Weber, B., Rangel, A., Wibral, M., & Falk, A. (2009). The medial prefrontal cortex exhibits money illusion. *Proceedings of the National Academy of Sciences, 106*(13), 5025-5028. https://doi.org/10.1073/pnas.0901490106

Mood-Congruent Judgment

Definition: Mood-congruent judgment is a cognitive bias where an individual's current emotional state influences their assessments of events or experiences. In other words, when we're feeling sunny, the world looks like a beach holiday; when we're blue, it's all thunderstorms.

Description: Just as your favorite playlist can set the mood for a road trip, your emotional state can cue up your judgments about the world around you. The mood-congruent judgment bias suggests that our current feelings color our perception and influence our judgments. Happiness can make us see the world through rose-tinted glasses, while sadness might make those same glasses look more like storm windows.

Examples: As we delve into the fascinating realms of religion, politics, and social issues, this bias weaves its colorful narrative.

In religion, mood-congruent judgment might shape how an individual interprets religious texts or experiences religious events. Feeling jubilant? That passage seems inspiring. Feeling somber? The very same passage takes on a melancholic tone.

In politics, a content voter might view a new policy as beneficial or promising, while a disgruntled one might see it as just another letdown.

In the context of social issues, someone who's feeling optimistic might perceive progress in the fight for equality, while another in a more pessimistic mood might focus on the remaining inequities.

Impact: Mood-congruent judgment can distort our perception of reality, leading to biased decisions and misinterpretations in various areas, including religion, politics, and social issues. This can foster discord and miscommunication, both on individual and societal levels.

Related Biases: This bias shares a dinner table with the affect heuristic (relying on current emotions to make decisions) and the optimism/pessimism bias (the tendency to expect the best/worst outcomes).

Countermeasures: Try to be aware of your current emotional state and its potential to influence your judgments. Seek diverse perspectives and objective information to balance your subjective mood-tinted view.

Key Takeaways: Our emotions can be like an Instagram filter for our judgments, coloring our perception of events and experiences. Recognizing this can help us adjust the exposure and make more accurate assessments. Life isn't always a beach holiday, but it doesn't have to be all thunderstorms, either.

References:

Mayer, J. D., Gaschke, Y. N., Braverman, D. L., & Evans, T. W. (1992). Mood-Congruent Judgment Is a General Effect. *Journal of Personality and Social Psychology, 63*(1), 119-132. https://doi.org/10.1037/0022-3514.63.1.119

Forgas, J. P. (1995). Mood and Judgment: The Affect Infusion Model (AIM). *Psychological Bulletin, 117*(1), 39–66. https://doi.org/10.1037/0033-2909.117.1.39

Bodenhausen, G. V., Kramer, G. P., & Süsser, K. (1994). Happiness and stereotypic thinking in social judgment. *Journal of Personality and Social Psychology, 66*(4), 621-632. https://doi.org/10.1037/0022-3514.66.4.621

Mood-Congruent Memory Bias

Definition: Mood-congruent memory bias refers to the cognitive bias where an individual's current emotional state influences the type of information they recall. Happy people remember happy things; sad people recall the blues. Life, it seems, has a soundtrack, and our memories dance to its tune.

Description: As though we're wearing mood-ring glasses, the hue of our emotions can tint the shade of our memories. This bias suggests that our current emotional state makes it easier for us to recall past events and experiences that were experienced in a similar emotional state. It's like your mind's personal DJ, spinning records that match your mood.

Examples: Just as a chameleon adapts to its environment, this bias pops up in varied shades across religion, politics, and social issues.

In religion, a joyful believer might remember more moments of spiritual elation, whereas a dispirited one might recall more instances of spiritual crisis or doubt.

In politics, a contented voter might recall more instances of governmental success, whereas an angry voter might remember more instances of scandal or mismanagement.

In social issues, an activist in a hopeful mood might remember past victories, whereas in a despairing mood, they might recall the challenges and setbacks more vividly.

Impact: Mood-congruent memory bias can subtly shape our personal narratives and worldviews, creating a skewed perspective of reality based on our emotional state. This might affect our understanding of historical events, personal relationships, and progress (or lack thereof) in key societal issues.

Related Biases: The mood-congruent memory bias is related to mood-congruent judgment bias (our mood influences our judgment of events) and the affective forecasting bias (predicting our future emotions based on our current ones).

Countermeasures: Increasing self-awareness of this bias and trying to ensure a balanced emotional state when recalling past events can help. Fact-checking our memories and seeking external perspectives can also provide a more objective view of the past.

Key Takeaways: Our current emotional state can DJ the playlist of our past, spinning records that match our mood. Recognizing this can help us ensure we're not only dancing to the sad songs or the happy hits, but getting a taste of the full album of our past.

References:

Bower, G. H. (1981). Mood and memory. *American Psychologist, 36*(2), 129-148. https://doi.org/10.1037/0003-066X.36.2.129

Eich, E., & Macaulay, D. (2000). Fundamental factors in mood-dependent memory. In J. P. Forgas (Ed.), *Feeling and thinking: The role of affect in social cognition* (pp. 109-130). Cambridge University Press.

Forgas, J. P., Laham, S. M., & Vargas, P. T. (2005). Mood effects on eyewitness memory: Affective influences on susceptibility to misinformation. *Journal of Experimental Social Psychology, 41*(6), 574-588. https://doi.org/10.1016/j.jesp.2004.11.005

Moral Credential Effect

Definition: The moral credential effect, much like a "get out of jail free" card in Monopoly, is our mind's way of using our past good deeds as justification for less than virtuous behavior now or in the future. It's the ol' "I did something good once, so I can slack off a bit now" mindset.

Description: Ever feel like your past good actions give you a free pass for being a bit naughty now? That's the moral credential effect hard at work, my friend. This cognitive bias convinces us that our previous ethical deeds have earned us some moral "brownie points" we can spend on indulging in behavior that's slightly less than saintly.

Examples: This bias has no boundaries and pops up in all aspects of life. In religion, after a week of rigorous fasting, you might feel entitled to gossip a bit. In politics, politicians often justify controversial policies by pointing to their previous good deeds. And in social issues, someone may feel that having advocated for climate change allows them to forgo recycling for a week. It's like a cognitive coupon system for good behavior, but the savings might not be as great as we think.

Impact: The moral credential effect can lead to a distorted self-image, ethical slippage, and an inconsistent approach to moral decision-making. In essence, it can turn us into cognitive cheapskates, always looking for a bargain in our moral choices.

Related Biases: The moral credential effect is part of the big happy family of self-serving biases, and it's particularly closely related to the self-licensing bias, where one action or choice provides mental permission for another.

Countermeasures: To balance our moral books, we need to recognize that our past deeds don't provide free passes for the future. Self-awareness and regular "moral audits" can keep us from falling into this trap. Consider it your personal moral accounting system.

Key Takeaways: Good deeds are not a renewable resource to be used for less ethical behavior later. Moral actions shouldn't be treated like a buy-one-get-one-free sale, where the second item might be a little shady. Remember, you can't bank ethics.

References:

Monin, B., & Miller, D. T. (2001). Moral credentials and the expression of prejudice. *Journal of Personality and Social Psychology, 81*(1), 33-43. https://doi.org/10.1037/0022-3514.81.1.33

Merritt, A. C., Effron, D. A., & Monin, B. (2010). Moral self-licensing: When being good frees us to be bad. *Social and Personality Psychology Compass, 4*(5), 344-357. https://doi.org/10.1111/j.1751-9004.2010.00263.x

Blanken, I., van de Ven, N., & Zeelenberg, M. (2015). A meta-analytic review of moral licensing. *Personality and Social Psychology Bulletin, 41*(4), 540-558. https://doi.org/10.1177/0146167215572134

Moral Luck

Definition: Moral luck is the cognitive bias that convinces us that the outcomes of our actions, often outside our control, affect the moral evaluation of the actions themselves. It's like karma hitting the jackpot on a slot machine of morality, irrespective of our actual intent.

Description: Imagine judging a reckless driver. If they make it home without incident, you might think they're merely foolhardy. But if they cause an accident, suddenly they're a menace to society. Same action, different outcomes, different moral judgments. That's moral luck playing with your moral compass.

Examples: In politics, a leader's decision might be hailed as strategic brilliance or derided as sheer incompetence, depending on outcomes they had little control over. In social issues, consider two protestors who throw rocks at a window. If one rock hits a passerby, that protestor is seen as more morally culpable, despite identical intent and action. And in religion, it might be the difference between a miracle and a tragic misfire of divine intervention.

Impact: The moral luck bias can lead to inconsistent and unfair moral judgments. It gives the illusion of a just world, where good outcomes reward good intentions and vice versa, ignoring the wild card of randomness.

Related Biases: Moral luck ties in with outcome bias, where judgments are skewed by the results rather than the decision-making process.

Countermeasures: To keep moral luck from clouding our judgments, we need to focus on actions and intentions rather than outcomes. A near-miss is as morally significant as a direct hit. And remember, even the best-laid plans of mice and men can be torpedoed by a wandering cat or a random meteorite.

Key Takeaways: Moral evaluations should be based on actions and intent, not the roll of the dice that life often is. When assessing morality, keep your eyes on the path, not where the dice fell.

References:

Nagel, T. (1979). Moral luck. *Mortal questions, 24*(2), 137-155. http://www.jstor.org/stable/2183688

Williams, B. (1981). Moral luck. Moral Luck: Philosophical Papers 1973–1980, 20, 35–56. https://doi.org/10.1017/CBO9781139165860.003

Hales, S. D. (2016). Why every theory of luck is wrong. *Nous, 50*(3), 490-508. https://doi.org/10.1111/nous.12089

Myside Bias

Definition: Myside bias, sometimes known as confirmation bias, is a cognitive bias where we cherry-pick information that supports our views and dismiss what doesn't. Essentially, it's our brain saying, "Who needs balanced judgment when you can just be right all the time?"

Description: This bias is driven by our tendency to favor information that bolsters our preexisting views while avoiding, dismissing, or twisting contradictory evidence. It's a cognitive Bermuda Triangle where opposing arguments mysteriously disappear.

Examples: Myside bias pops up like unwanted advertisements in social issues, religion, and politics. Take climate change, for example. You might find yourself scoffing at snowfall as evidence against global warming or ignoring scientific data that doesn't align with your belief. In politics, myside bias turns debates into cheerleading contests, where the loudest voice is often confused with the most correct one.

Impact: This bias limits our understanding of the world, promotes polarization, and undermines productive discourse. It's like living in an echo chamber where you're the superstar, and all the echoes are your adoring fans. While gratifying, it's not exactly the recipe for intellectual growth or informed decision-making.

Related Biases: Myside bias is the charismatic leader of a dysfunctional band of biases, including the confirmation bias, belief bias, and selective perception.

Countermeasures: To curb this bias, expose yourself to a variety of viewpoints. It might be like a cold shower on a winter morning, but it's vital for a balanced perspective. Practicing critical thinking and encouraging open-minded discussions can also help.

Key Takeaways: Myside bias can turn us into one-man cheering squads for our own beliefs. To escape this echo chamber, we should embrace differing opinions and question our own biases. Because let's face it, being right all the time can get pretty boring.

References:

Stanovich, K. E., West, R. F., & Toplak, M. E. (2013). Myside bias, rational thinking, and intelligence. *Current Directions in Psychological Science, 22*(4), 259-264. https://doi.org/10.1177/0963721413480174

Mercer, J. (2005). Prospect theory and political science. *Annual Review of Political Science, 8*, 1-21. https://doi.org/10.1146/annurev.polisci.8.082103.104911

Sperber, D., Clément, F., Heintz, C., Mascaro, O., Mercier, H., Origgi, G., & Wilson, D. (2010). Epistemic vigilance. *Mind & Language, 25*(4), 359-393. https://doi.org/10.1111/j.1468-0017.2010.01394.x

Naive Realism

Definition: Naive realism, or the "I'm right, you're biased" syndrome, is the conviction that we see the world as it is, while others are influenced by their biases and are thus less objective. It's the mental equivalent of claiming, "I don't have an accent; everyone else does."

Description: This bias stems from our assumption that our perceptions and interpretations of events are the unvarnished truth. We firmly believe our view is the most accurate, like we're all wearing a pair of magic spectacles that render the world as it really is.

Examples: Consider a controversial political issue. You might think your stance is grounded in pure facts and logic, while those on the other side are merely brainwashed or misinformed. This isn't limited to politics. Whether it's a feud over who makes the best pizza or disagreements on religious doctrines, naive realism is always ready to assure us that we're the beacon of objectivity in a sea of bias.

Impact: Naive realism promotes overconfidence in our own beliefs and contributes to polarization and misunderstanding. It's like throwing gasoline on the fires of disagreement, turning civil debates into flame wars.

Related Biases: This bias is the head honcho of a gang that includes the bias blind spot, false consensus effect, and the fundamental attribution error.

Countermeasures: To mitigate this bias, practice intellectual humility. Accept that your perspective is just that – a perspective, not absolute reality. Encourage empathy, listen to differing viewpoints, and keep in mind that others may view the world differently without being biased or wrong.

Key Takeaways: Naive realism is the cognitive equivalent of insisting that you're the center of the universe. To avoid its pitfalls, it's important to remember that everyone has their own unique viewpoint, and that's okay. After all, a world where everyone agrees with you would be as dull as a never-ending game of solitaire.

References:

Ross, L., & Ward, A. (1996). Naive realism in everyday life: Implications for social conflict and misunderstanding. In T. Brown, E. S. Reed, & E. Turiel (Eds.), *Values and knowledge* (pp. 103-135). Erlbaum.

Pronin, E., Gilovich, T., & Ross, L. (2004). Objectivity in the eye of the beholder: Divergent perceptions of bias in self versus others. *Psychological Review, 111*(3), 781-799. https://doi.org/10.1037/0033-295X.111.3.781

Robinson, R. J., Keltner, D., Ward, A., & Ross, L. (1995). Actual versus assumed differences in construal: "Naive realism" in intergroup perception and conflict. *Journal of Personality and Social Psychology, 68*(3), 404-417. https://doi.org/10.1037/0022-3514.68.3.404

Narrative Fallacy

Definition: The narrative fallacy is the cognitive bias that compels us to squeeze complex realities into simplistic, cohesive stories. It's our brain's endearing yet misguided attempt to make the chaos of life resemble a neatly written novel.

Description: We humans love a good story. So much so that we impose narrative structures on events to make sense of them. While this makes for entertaining campfire tales, it often leads us to overlook nuances and complexities, misinterpret randomness, and assign causality where none exists.

Examples: Think of a political campaign where the candidate's life story is neatly packaged into an underdog-overcoming-all-odds tale. Or consider how conspiracy theories often provide a seductively simplistic narrative for complex global events, making them enticing despite their lack of factual grounding.

Impact: The narrative fallacy can lead us to erroneous conclusions and unverified assumptions. It can make us susceptible to propaganda, misinformation, and conspiracy theories, effectively turning us into putty in the hands of a good storyteller.

Related Biases: Its partners in crime include the confirmation bias, representativeness heuristic, and the illusion of causality.

Countermeasures: To combat this bias, embrace complexity. It's okay if life doesn't resemble a best-selling thriller. Critical thinking, data-driven decisions, and skepticism towards overly simplistic narratives can help keep this bias at bay.

Key Takeaways: Life is not a book, and sometimes there's no grand storyline or satisfying resolution. Accepting this reality can protect us from the pitfalls of the narrative fallacy.

References:

Taleb, N. N. (2007). The Black Swan: The Impact of the Highly Improbable. New York: Random House.

Kahneman, D., & Tversky, A. (1972). Subjective probability: A judgment of representativeness. Cognitive Psychology, 3(3), 430–454. https://doi.org/10.1016/0010-0285(72)90016-3

Matute, H., Yarritu, I., & Vadillo, M. A. (2011). Illusions of causality at the heart of pseudoscience. British Journal of Psychology, 102(3), 392–405. https://doi.org/10.1111/j.2044-8295.2011.02008.x

Negativity Bias

Definition: Negativity bias is the mental equivalent of a pesky mosquito at a summer barbecue, where we tend to focus more on negative experiences than positive ones. It's the cognitive bias that makes us more inclined to recall, react to, and be influenced by unfavorable information.

Description: Negativity bias stems from our evolutionary need for survival. From an evolutionary perspective, focusing on threats (negative experiences) rather than rewards (positive experiences) would have helped our ancestors stay alive. Nowadays, it may not be as useful when the biggest threat to our day is a negative comment on social media, but hey, evolution takes time!

Examples: When looking at online reviews, are you not inexplicably drawn to the one-star ones like a moth to a flame? That's negativity bias. Or consider how a minor disagreement with your significant other can eclipse the dozen affectionate moments you shared that day.

Impact: On the societal level, negativity bias can fuel fear-based politics and sensationalist media narratives. On the personal level, it can lead to unnecessary stress, hindered decision-making, and even mental health problems such as anxiety and depression.

Related Biases: The negativity bias is closely related to loss aversion, where losses are felt more strongly than gains of the same magnitude, and the availability heuristic, which involves relying on immediate examples when evaluating a specific topic or decision.

Countermeasures: Implement strategies like mindfulness, cognitive behavioral techniques, or a good old-fashioned gratitude journal to shift the focus towards positive experiences. News detox, anyone?

Key Takeaways: Negativity bias is a leftover from our cave-dwelling days that can sour our modern lives. Being aware of this bias and consciously fostering positivity can help bring balance.

References:

Rozin, P., & Royzman, E. B. (2001). Negativity bias, negativity dominance, and contagion. Personality and Social Psychology Review, 5(4), 296–320. https://doi.org/10.1207/S15327957PSPR0504_2

Kahneman, D., & Tversky, A. (1979). Prospect theory: An analysis of decision under risk. Econometrica, 47(2), 263-291. https://doi.org/10.2307/1914185

Tversky, A., & Kahneman, D. (1973). Availability: A heuristic for judging frequency and probability. Cognitive Psychology, 5(2), 207-232. https://doi.org/10.1016/0010-0285(73)90033-9

Neglect of Duration

Definition: Neglect of duration is the cognitive bias equivalent of "time flies when you're having fun." It refers to our tendency to disregard the duration of events when assessing their value or impact, making a mountain out of a molehill for short but intense experiences, while the long, less intense ones become the forgotten marathon runners.

Description: The exact causes of neglect of duration are still under debate, but it could be attributed to our limited attentional resources or an evolutionary remnant favoring immediate threats and rewards. Ironically, in our haste to make quick decisions, we end up missing the long game.

Examples: Let's take a political rally. A charismatic speaker gives a fiery 15-minute speech that leaves the crowd buzzing. The same crowd endures two hours of mediocre speeches without blinking. Guess who gets the headlines the next day?

Impact: Neglect of duration can have profound implications in policy-making, healthcare, and personal decision-making. For example, in healthcare, a patient might opt for a treatment with severe but short-lived side effects over a milder, longer-term treatment.

Related Biases: This bias is related to the peak-end rule, where we judge experiences based on how they were at their peak and their end, not the duration. The duration neglect is also a variant of the more general attentional bias, where we pay more attention to certain things while simultaneously ignoring others.

Countermeasures: A good start is simply being aware of the bias and intentionally considering the duration of events when making judgments or decisions. For those of us who struggle with this, setting reminders or using decision-making tools might help.

Key Takeaways: The neglect of duration bias may lead us to overemphasize the short and intense while downplaying the long and sustained. Remember, life isn't a sprint; it's a marathon. Time to start thinking like a long-distance runner!

References:

Kahneman, D., Fredrickson, B. L., Schreiber, C. A., & Redelmeier, D. A. (1993). When more pain is preferred to less: Adding a better end. Psychological Science, 4(6), 401–405. https://doi.org/10.1111/j.1467-9280.1993.tb00589.x

Hsee, C. K., & Zhang, J. (2004). Distinction bias: Misprediction and mischoice due to joint evaluation. Journal of Personality and Social Psychology, 86(5), 680–695. https://doi.org/10.1037/0022-3514.86.5.680

Neglect of Probability

Definition: Neglect of Probability is the cognitive bias in which we, humans, view the world not as a roulette wheel, but more like a comic book, neglecting the role of chance and probability in the outcomes we see or expect.

Description: Whether it's due to poor understanding of statistics, the allure of deterministic explanations, or just our brains taking a shortcut, neglect of probability can lead us to make questionable decisions. Like the time you bought 50 lottery tickets because you were 'feeling lucky', ignoring the 1 in 14 million odds.

Examples: In politics, pundits often neglect probability when they unequivocally declare the winner of the next election based on a single poll. Or consider the heated debates on social issues where anecdotal evidence is given more weight than large-scale statistical data.

Impact: Neglect of probability can lead to significant distortions in decision-making, both at an individual and societal level. It can influence our perceptions of risk, causing us to undervalue precautions or overestimate the likelihood of rare events.

Related Biases: This bias often operates in tandem with the Availability Heuristic, which makes us overestimate the probability of events that come easily to mind. It also ties in with the Gambler's Fallacy, where people believe past events influence future probabilities.

Countermeasures: Education about basic statistical concepts and frequent exposure to probabilistic thinking can help. Also, try replacing "it will happen" in your thoughts with "there's a chance it might happen."

Key Takeaways: While neglect of probability might make for compelling narratives, it's not the best guide for decision-making. Remember, life isn't a comic book; it's more like roulette. Don't ignore the spinning wheel!

References:

Tversky, A., & Kahneman, D. (1974). Judgment under Uncertainty: Heuristics and Biases. Science, 185(4157), 1124-1131. https://doi.org/10.1126/science.185.4157.1124

Kahneman, D., & Tversky, A. (1982). The psychology of preferences. Scientific American, 246(1), 160-173. https://doi.org/10.1038/scientificamerican0182-160

Normalcy Bias

Definition: Normalcy Bias is the cognitive propensity to believe that things will always function the way they normally have, and therefore underestimate both the possibility and the effects of a disaster or drastic change. It's essentially our mind's own version of a 'this too shall pass' meme.

Description: This bias manifests in our tendency to dismiss dire warnings and catastrophes, underestimating their seriousness or probability. It's the mental equivalent of sticking our fingers in our ears and humming a happy tune when confronted with disturbing news.

Examples: On a large scale, normalcy bias is evident in how societies respond to climate change warnings. Many people assume things will continue as they always have, despite mounting evidence to the contrary. Similarly, in politics, normalcy bias can lead to the neglect of systemic issues, as people assume "it's always been this way."

Impact: The consequences of this bias can be significant. By minimizing the likelihood of negative outcomes, we can fail to take timely action or adequately prepare for change, from saving for a rainy day to preparing for a pandemic.

Related Biases: The Optimism Bias, which leads people to believe they are less at risk of experiencing a negative event compared to others, and the Status Quo Bias, where people prefer things to stay the same, often intertwine with the Normalcy Bias.

Countermeasures: Developing a more realistic outlook, and a willingness to face uncomfortable truths can mitigate this bias. Scenario planning and risk assessments are also valuable tools for countering the effects of normalcy bias in decision-making.

Key Takeaways: The normalcy bias provides comforting denial in the face of change or threat. But remember, just because you're humming a happy tune, doesn't mean the music can't change. Recognizing this bias can help us better prepare for the unexpected.

References:

Helweg-Larsen, M. (1999). (The lack of) optimistic biases in response to the 1994 Northridge earthquake: the role of personal experience. Basic and Applied Social Psychology, 21(2), 119-129. https://doi.org/10.1207/s15324834basp2102_3

Malmendier, U., & Nagel, S. (2011). Depression Babies: Do Macroeconomic Experiences Affect Risk-Taking? The Quarterly Journal of Economics, 126(1), 373-416. https://doi.org/10.1093/qje/qjq004

Not Invented Here (NIH) Syndrome

Definition: Not Invented Here (NIH) Syndrome is a cognitive bias that results in an aversion to ideas, solutions, or products that originate from outside a particular group or organization. This bias effectively acts as an internal bouncer, only allowing in home-grown ideas while giving foreign concepts a rough bounce to the curb.

Description: NIH Syndrome often stems from a misplaced sense of pride in self-reliance, protectionism, or the need to maintain control. It can result in rejecting perfectly good ideas or solutions simply because they were not home-brewed.

Examples: The NIH Syndrome can rear its head in a political setting when politicians refuse to adopt policies successful in other countries, simply because they were not "invented" domestically. It's like refusing to use a Swedish flat-pack furniture because it wasn't designed by your Uncle Bob who makes everything with duct tape.

Impact: The impact of NIH Syndrome can be substantial, leading to suboptimal outcomes, stifled innovation, and wasted resources. For organizations, it can even mean the difference between success and failure.

Related Biases: NIH Syndrome is a close cousin to Confirmation Bias, where one favors information that confirms pre-existing beliefs and In-group Bias, where one shows a preference for those within their own group.

Countermeasures: Encouraging diversity of thought, embracing change, and promoting a culture of openness can help combat NIH Syndrome. Critical thinking and evidence-based decision making can also help to ensure the best ideas are accepted, regardless of their origin.

Key Takeaways: Not Invented Here Syndrome can stifle innovation and lead to missed opportunities. Remember, just because an idea was born in a different zip code, doesn't mean it can't become a fully naturalized citizen of your mind.

References:

Katz, R., & Allen, T. J. (1982). Investigating the Not Invented Here (NIH) syndrome: A look at the performance, tenure, and communication patterns of 50 R & D Project Groups. R&D Management, 12(1), 7-20. https://doi.org/10.1111/j.1467-9310.1982.tb00478.x

Lichtenthaler, U., & Ernst, H. (2006). Attitudes to externally organising knowledge management tasks: a review, reconsideration and extension of the NIH syndrome. R&D Management, 36(4), 367-386. https://doi.org/10.1111/j.1467-9310.2006.00451.x

Observer Expectancy Effect

Definition: The Observer Expectancy Effect is a cognitive bias where researchers' expectations unconsciously influence the outcome of their study. It's like a self-fulfilling prophecy with a PhD.

Description: This bias takes root when the observer subtly communicates their expectations to the subjects, inadvertently influencing their behavior or responses. The observer then notes this 'anticipated' behavior, essentially seeing what they want to see.

Examples: Imagine a sociologist conducting a survey on political inclinations. If they wear a button of their favorite candidate while conducting interviews, respondents may adjust their responses due to the apparent bias. It's the equivalent of asking someone about their diet while munching on a bag of chips.

Impact: The Observer Expectancy Effect can lead to skewed data, incorrect conclusions, and potentially harmful decisions based on that data. In a broader sense, it may foster and perpetuate biases in fields like social sciences, medical research, and yes, politics.

Related Biases: This bias is a sibling to Confirmation Bias, where one tends to favor information that confirms their pre-existing beliefs. It's also closely related to the Experimenter's Bias and Hawthorne Effect.

Countermeasures: Blind and double-blind experimental designs can help curb this bias. Additionally, cultivating awareness of personal biases and following rigorous methodologies can help ensure research integrity.

Key Takeaways: In scientific research and decision making, your expectations can subtly contort reality. It's vital to take steps to minimize the influence of your own biases on the outcome.

References:

Rosenthal, R., & Fode, K. L. (1963). The effect of experimenter bias on the performance of the albino rat. Behavioral Science, 8(3), 183-189. https://doi.org/10.1002/bs.3830080302

Orne, M. T. (1962). On the social psychology of the psychological experiment: With particular reference to demand characteristics and their implications. American psychologist, 17(11), 776. https://doi.org/10.1037/h0043424

Observational Selection Bias

Definition: Observational Selection Bias is a cognitive wrinkle where we suddenly start noticing things we've recently learned about, assuming that the frequency of these things has increased. It's the universe's version of "when you buy a new car, suddenly you see the same model everywhere."

Description: It's not that these things are appearing more frequently; they were always there. It's just that our brain has been tuned to notice them now. This bias is like getting a VIP backstage pass that lets you see the details you've been missing.

Examples: If you just learned about a peculiar fashion trend like wearing socks with sandals (why, though?), you might start noticing this audacious style all over town. In politics, if you recently read about gerrymandering, you might suddenly notice numerous articles, discussions, and posts about it.

Impact: This bias can make us overestimate the prevalence of an event, leading to incorrect assessments and, often, hilariously wrong conclusions. For instance, you might believe that socks with sandals are the latest haute couture, or that gerrymandering is a new political tactic, rather than an unfortunate tradition.

Related Biases: The Baader-Meinhof Phenomenon, or frequency illusion, is a close cousin. This bias is also related to Confirmation Bias, as our brain loves to see more evidence that supports our beliefs.

Countermeasures: Awareness is your superpower here. Recognize that noticing something more often doesn't necessarily mean it's occurring more often. Statistics and reliable data can also help to provide a more balanced perspective.

Key Takeaways: Just because you notice something more doesn't mean it's happening more. Your brain is simply spotlighting the new kid on the block.

References:

Krippendorff, K. (2018). Content analysis: An introduction to its methodology. Sage publications.

Tunney, R. J., & Ziegler, F. V. (2015). Surrogate utility estimation by long-term partners and unfamiliar dyads. Frontiers in psychology, 6, 315. https://doi.org/10.3389/fpsyg.2015.00315

Omission Bias

Definition: The Omission Bias is our quirky tendency to judge harmful actions as worse than equally harmful omissions or inactions. It's the cognitive equivalent of "I didn't start the fire; I just didn't put it out."

Description: People have a tendency to see harmful acts as worse than harmful omissions, even if the outcomes are the same or worse. We're more comfortable with bad things that happen because we *didn't* act than bad things that happen because we *did*. Like an aversion to action, this bias helps us evade responsibility.

Examples: Suppose a politician decides not to intervene in a humanitarian crisis in another country, resulting in numerous casualties. Many may still view this politician more favorably than one who engages in an intervention that results in similar casualties. Or consider parents who decide not to vaccinate their children, judging the potential harm from action (vaccination side effects) as worse than potential harm from inaction (contracting a preventable disease).

Impact: Omission bias can lead to harmful inaction in situations where action is needed. It can also contribute to blaming victims of circumstances and abdicating responsibility for problems.

Related Biases: The Bystander Effect, where people are less likely to help a victim when others are present, and Status Quo Bias, a preference for keeping things the same by doing nothing, are kindred spirits.

Countermeasures: To counter this bias, we should assess the potential outcomes of both action and inaction, not just focus on the potential risks of taking action. In other words, sometimes, it's not just the thought that counts.

Key Takeaways: Just because you didn't do anything doesn't mean you didn't do anything wrong. Both actions and inactions can have consequences, and both can be judged.

References:

Baron, J., & Ritov, I. (2004). Omission bias, individual differences, and normality. Organizational Behavior and Human Decision Processes, 94(2), 74–85. https://doi.org/10.1016/j.obhdp.2004.03.003

Cushman, F., Murray, D., Gordon-McKeon, S., Wharton, S., & Greene, J. D. (2012). Judgment before principle: engagement of the frontoparietal control network in condemning harms of omission. Social cognitive and affective neuroscience, 7(8), 888–895. https://doi.org/10.1093/scan/nsr072

Optimism Bias

Definition: The Optimism Bias, or "our collective unwarranted cheeriness," is our tendency to overestimate the likelihood of positive events and underestimate the likelihood of negative events in our lives.

Description: Our brains have an inbuilt Pollyanna mechanism that makes us believe the future will be better than the past, and that we'll likely dodge life's metaphorical bullets. It's like living in a perpetual state of "the glass is half-full," even when it's clearly dribbling down the side.

Examples: For instance, we tend to believe that we're less likely than others to suffer from medical conditions, experience a divorce, or become the victim of a crime. In politics, it might manifest in believing our candidate of choice will surely win an election, despite polls indicating otherwise.

Impact: While optimism can be beneficial for motivation and mental health, it can also lead to poor decision-making and lack of preparedness. In social issues, excessive optimism can lead to complacency in the face of systemic problems, while in politics, it might blind us to the possibility of an undesirable outcome.

Related Biases: The overconfidence effect, where we overestimate our abilities, and the planning fallacy, our tendency to underestimate the time needed to complete tasks, are siblings in the family of rosy misconceptions.

Countermeasures: Being aware of our optimism bias can help. Reality-checking our expectations and considering potential obstacles can also aid in keeping this bias at bay.

It's good to be positive, but not to the point of ignoring the thunderstorm while planning a picnic.

Key Takeaways: Although optimism can be a great asset, unchecked optimism can cloud judgment. It's important to balance positivity with a healthy dose of reality. And remember, expecting the best doesn't prevent the worst.

References:

Sharot, T. (2011). The optimism bias. Current Biology, 21(23), R941–R945. https://doi.org/10.1016/j.cub.2011.10.030

Weinstein, N. D. (1980). Unrealistic optimism about future life events. Journal of Personality and Social Psychology, 39(5), 806–820. https://doi.org/10.1037/0022-3514.39.5.806

Outcome Bias

Definition: Outcome Bias is our judgment-skewing friend that convinces us to evaluate a decision based on its outcome, rather than the quality of the decision at the time it was made.

Description: Imagine a psychic foresaw that bungee jumping without a cord was a great idea - because a massive pile of marshmallows had miraculously appeared beneath the cliff. After taking the leap and miraculously surviving thanks to the fluffy white mound, we shouldn't conclude that jumping without a cord was a wise decision. But that's exactly what outcome bias does.

Examples: In the political realm, a leader might make a highly risky decision that luckily results in positive outcomes, leading us to see the decision as smart rather than reckless. In religious debates, a successful but dangerously unprepared missionary expedition might be viewed as "guided by divine providence," rather than "rash and irresponsible."

Impact: Outcome Bias can lead to flawed decision-making and dangerous risk-taking, by convincing us that as long as things work out in the end, the decision was good. This can perpetuate poor decision-making strategies in the long run.

Related Biases: Its cousin, the Hindsight Bias, can exacerbate the Outcome Bias by making us believe we "knew it all along" when an outcome is revealed.

Countermeasures: Awareness of this bias can help us scrutinize decisions more thoroughly, independently of their outcomes. Analyzing decisions based on the information and circumstances at the time they were made can also keep this bias at bay.

Key Takeaways: A good outcome doesn't always mean a good decision was made, and vice versa. Don't let the sweet lure of successful outcomes cloud your judgment of the decisions that led there.

References:

Baron, J., & Hershey, J. C. (1988). Outcome bias in decision evaluation. Journal of Personality and Social Psychology, 54(4), 569–579. https://doi.org/10.1037/0022-3514.54.4.569

Caputo, A., & Trivella, P. (2020). Outcome bias in decision making. Journal of Business Research, 109, 278-287. https://doi.org/10.1016/j.jbusres.2019.11.040

Outgroup Homogeneity Bias

Definition: The Outgroup Homogeneity Bias is the cognitive slip-up that tricks us into thinking members of groups to which we don't belong (outgroups) are more similar to each other than they actually are. It's like looking at a bag of mixed nuts and deciding all the peanuts are identical because they're not cashews.

Description: This bias takes root in our mind garden due to a lack of exposure or familiarity with the individuals in the outgroup. It's often easier to see the nuances and individual differences within our own group (the ingroup), because, let's face it, we've spent more time with them.

Examples: This bias is seen in politics when people generalize all Democrats or all Republicans based on the actions or beliefs of a few. In religion, someone might assume all followers of a certain faith hold identical beliefs, overlooking the diversity within the group. The same can occur within societal contexts, such as believing all millennials are avocado-toast-devouring, tech-savvy folks.

Impact: The Outgroup Homogeneity Bias can foster stereotypes, prejudice, and discrimination, reinforcing the walls of misunderstanding and mistrust between different groups.

Related Biases: The Ingroup Bias, which is the tendency to favor one's own group, often dances in unison with the Outgroup Homogeneity Bias. They're like two peas in a bias pod.

Countermeasures: Increase exposure to and interaction with diverse groups. The more you learn about the 'other,' the less they remain an 'other.' Embrace diversity, and encourage others to do the same.

Key Takeaways: No group is a monolith. Just as every cashew in your mixed nuts bag is unique, so are individuals within a group. Expand your horizons, shatter the assumptions, and appreciate the diversity.

References:

Park, B., & Judd, C. M. (1990). Measures and models of perceived group variability. Journal of Personality and Social Psychology, 59(2), 173–191. https://doi.org/10.1037/0022-3514.59.2.173

Ostrom, T. M., & Sedikides, C. (1992). Out-group homogeneity effects in natural and minimal groups. Psychological Bulletin, 112(3), 536–552. https://doi.org/10.1037/0033-2909.112.3.536

Overconfidence Bias

Definition: Overconfidence Bias is the charming fellow that convinces us we're more capable, knowledgeable, or accurate in our judgments than we actually are. It's like being certain you're the next Mozart when you can barely manage a decent rendition of "Chopsticks" on the piano.

Description: This cognitive distortion thrives on our desire to see ourselves positively, coupled with the limited feedback about our true abilities. It's the ego's pet, feeding on our lack of self-awareness, until it grows into a full-fledged illusion of competence.

Examples: Overconfidence Bias can manifest in religious fervor, with individuals asserting that their interpretation of sacred texts is the only correct one. In politics, it can lead candidates to overestimate their popularity, believing they have an election in the bag when they're actually trailing. In social issues, it can result in people claiming to understand complex problems without having all the facts.

Impact: Overconfidence can lead to poor decision-making, increased risk-taking, and a lack of receptiveness to opposing viewpoints. Let's face it, no one likes to admit they're wrong, especially when they're certain they're right. The resulting miscalculations can ripple through everything from political elections to social movements.

Related Biases: Dunning-Kruger Effect is a close cousin. It describes a situation where people with low ability at a task overestimate their ability, like a tone-deaf singer convinced they're destined for Broadway.

Countermeasures: Foster humility, invite feedback, and maintain an attitude of lifelong learning. Remember, it's okay not to know everything, as long as you're willing to learn.

Key Takeaways: The mind can be a bit of a flatterer, often convincing us we're more competent than we actually are. To keep this bias in check, practice humility, invite critique, and always stay open to learning.

References:

Moore, D. A., & Healy, P. J. (2008). The trouble with overconfidence. Psychological Review, 115(2), 502–517. https://doi.org/10.1037/0033-295X.115.2.502

Kruger, J., & Dunning, D. (1999). Unskilled and unaware of it: how difficulties in recognizing one's own incompetence lead to inflated self-assessments. Journal of Personality and Social Psychology, 77(6), 1121–1134. https://doi.org/10.1037/0022-3514.77.6.1121

Overestimation Bias

Definition: Overestimation Bias is the cognitive equivalent of an overzealous cheerleader, leading us to overvalue the likelihood, importance, or size of something. It's like believing that your chances of winning the lottery are akin to getting struck by lightning, when in reality, the latter is far more probable.

Description: This bias thrives in the gray area between optimism and reality, where we inflate probabilities or significance based on our preferences or fears. It's like watching a horror movie at midnight and deciding every creak in the house is a chainsaw-wielding maniac.

Examples: In religion, this can lead to an overestimation of the severity of divine punishment, causing undue anxiety. In politics, it can cause individuals to exaggerate the importance of a single vote in a national election. For social issues, overestimation bias can lead to an overinflation of the impact of a single event or individual.

Impact: Overestimation can lead to skewed decision-making, creating undue stress, and fostering an unrealistic worldview. It can distort perceptions in everything from political discourse to social policy debates, often leading to more heat than light.

Related Biases: Availability Bias, where we estimate the probability of an event based on how easily examples come to mind, is a related bias. If you can easily recall instances of a particular event, you might overestimate its frequency or importance.

Countermeasures: Apply the principle of Occam's Razor (the simplest explanation is usually the right one), promote critical thinking, and always ask for evidence.

Key Takeaways: Overestimation bias can lead us to inflate probabilities and magnify importance. Keep this bias at bay by promoting critical thinking, seeking evidence, and favoring simplicity when contemplating probabilities.

References:

Tversky, A., & Kahneman, D. (1973). Availability: A heuristic for judging frequency and probability. Cognitive Psychology, 5(2), 207-232. https://doi.org/10.1016/0010-0285(73)90033-9

Moore, D. A., & Small, D. A. (2007). Error and bias in comparative judgment: On being both better and worse than we think we are. Journal of Personality and Social Psychology, 92(6), 972–989. https://doi.org/10.1037/0022-3514.92.6.972

Overprecision

Definition: Overprecision is that cocky cousin of overconfidence, where we not only think we're right but believe we're precisely right. Like deciding the precise number of jelly beans in a jar from just a quick glance. Spoiler alert: It's usually not accurate.

Description: This cognitive bias reflects our tendency to be excessively confident in the precision or accuracy of our beliefs and predictions. It's a biased self-assessment where we underestimate the range of potential outcomes or overestimate the exactness of our knowledge.

Examples: An example of overprecision in religion might be when a person insists that a particular interpretation of a religious text is the one and only correct one, disregarding other possible interpretations. In politics, it can manifest in forecasts of election outcomes, where analysts present precise predictions without acknowledging the inherent uncertainty. And in social issues, it can be seen in the overprecise estimations about the outcomes of policy changes, without sufficient data or consideration of multiple factors involved.

Impact: Overprecision can lead to miscalculations and poor decision-making as it undermines the natural uncertainty inherent in most situations. It can cause polarization in religious and political discussions and lead to potentially harmful policies in social issues when the complex dynamics involved are not fully appreciated.

Related Biases: Overconfidence Bias, as it involves an excessive belief in one's own abilities or correctness, is a related bias.

Countermeasures: To mitigate the effects of overprecision, one could apply statistical reasoning, seek external validation, and actively consider alternative outcomes or interpretations. Emphasize the natural variability and uncertainty inherent in most situations.

Key Takeaways: Overprecision is the tendency to be excessively confident in the precision of our beliefs and predictions. It can lead to miscalculations and misunderstandings, but can be mitigated by statistical reasoning and an openness to alternatives.

References:

Moore, D. A., & Healy, P. J. (2008). The trouble with overconfidence. Psychological Review, 115(2), 502–517. https://doi.org/10.1037/0033-295X.115.2.502

Malmendier, U., & Taylor, D. (2015). On the verges of overconfidence. Journal of Economic Perspectives, 29(4), 3-8. https://doi.org/10.1257/jep.29.4.3

Overreaction Bias

Definition: Overreaction bias is when we transform from calm and composed individuals into human-sized, emotion-charged pendulums, swinging wildly from one extreme to another based on a single piece of news or event. It's like finding a single ant in your kitchen and immediately calling for the full fumigation of your home.

Description: This cognitive bias refers to our tendency to respond more intensely to information or events than warranted, particularly when that information contradicts our beliefs or expectations. The overreaction bias often results from our failure to adequately consider the reliability and representativeness of information before reacting to it.

Examples: In the realm of religion, overreaction bias could manifest in a person losing faith after a single negative experience with a religious institution. Politically, a minor policy mishap can lead to an immediate and severe drop in approval ratings, thanks to this bias. For social issues, overreaction bias might be seen when people jump to drastic conclusions based on a single incident, leading to hasty and potentially unfair judgments.

Impact: Overreaction bias can cause significant disruptions in decision-making and lead to ill-informed and impulsive choices. In religion, politics, and social issues, it can contribute to increased polarization and unnecessary conflict.

Related Biases: This bias is closely related to Recency Bias, where the most recent information or event disproportionately influences our decisions and judgments, causing us to overreact.

Countermeasures: The best remedy for overreaction bias is to take a pause, breathe, and objectively evaluate the situation. Seek a broad range of information before making decisions and consider the context and reliability of the information at hand. Emotional regulation techniques can also be helpful.

Key Takeaways: Overreaction bias is when our emotional reaction overshoots the appropriate response to a situation. By being aware of this bias, seeking varied sources of information, and practicing emotional regulation, we can make more balanced and rational decisions.

References:

De Bondt, W. F., & Thaler, R. H. (1985). Does the Stock Market Overreact? The Journal of Finance, 40(3), 793-805. https://doi.org/10.1111/j.1540-6261.1985.tb05004.x

Langer, E. J., & Roth, J. (1975). Heads I win, tails it's chance: The illusion of control as a function of the sequence of outcomes in a purely chance task. Journal of Personality and Social Psychology, 32(6), 951–955. https://doi.org/10.1037/0022-3514.32.6.951

Overtrust

Definition: Overtrust is a cognitive bias where we misplace our faith in others with the audacity of a teenager borrowing their parent's car for the first time. It refers to our tendency to trust others' intentions or abilities more than is warranted by the situation or evidence at hand.

Description: Overtrust is primarily driven by our innate need to believe in others and maintain social harmony. This cognitive bias involves overestimating the honesty, reliability, or competence of others, often leading to an overly optimistic assessment of their intentions or abilities. A dollop of wishful thinking and a dash of naiveté add up to make this bias a prevalent phenomenon.

Examples: A classic example in religion could be the blind trust in a spiritual leader despite inconsistencies or controversies. Politically, overtrust is seen when voters believe campaign promises with little scrutiny. On social issues, overtrust can manifest when people take statements at face value without questioning the source's credibility or bias.

Impact: Overtrust can lead to manipulation, deception, or disappointment when our trust is unmet. It can make us vulnerable to scams or misleading information and can fuel conflicts and disillusionment in various domains, including religion, politics, and social issues.

Related Biases: This bias is closely related to the Authority Bias, where we tend to trust and obey figures of authority without question, and Confirmation Bias, where we favor information that confirms our pre-existing beliefs.

Countermeasures: Practice skepticism and critical thinking. Verify information before accepting it, and hold others accountable for their actions. Emotional intelligence can also help discern the sincerity and capability of others.

Key Takeaways: Overtrust reflects our tendency to overestimate the trustworthiness and competence of others. By cultivating a healthy dose of skepticism and holding others accountable, we can protect ourselves from the potential pitfalls of this bias.

References:

Barber, B., & Odean, T. (2001). Boys will be boys: Gender, overconfidence, and common stock investment. The Quarterly Journal of Economics, 116(1), 261-292. https://doi.org/10.1162/003355301556400

Bonaccio, S., & Dalal, R. S. (2006). Advice taking and decision-making: An integrative literature review, and implications for the organizational sciences. Organizational Behavior and Human Decision Processes, 101(2), 127-151. https://doi.org/10.1016/j.obhdp.2006.07.001

Pareidolia

Definition: Pareidolia, my friend, is seeing faces in clouds or Elvis in a slice of toast. It's a psychological phenomenon where we interpret vague and random stimuli as having significant, often specific and identifiable, meanings.

Description: Pareidolia comes from our brain's incredible, yet sometimes overzealous, pattern-recognition abilities. The brain is an incredible problem-solving machine that occasionally overdoes it, like a DJ with the bass at 3 am. It's so eager to find patterns, sometimes it sees them where none exist. That's pareidolia – perceiving significant images in random or unrelated objects or patterns.

Examples: Religiously, it's seen when people perceive divine figures in everyday objects. In politics, one may 'read between the lines' of a politician's speech to find hidden meanings that might not exist. Socially, we see it in claims of ghost sightings in blurry photos or hearing secret messages in songs played backwards.

Impact: Pareidolia can affect our perceptions and beliefs, leading to misinterpretations or conspiracy theories. While mostly harmless (except maybe to the burnt toast), it can fuel misinformation and superstition. It's also a favorite device of illusionists and can make us fall for 'fake news' faster than a cat video goes viral.

Related Biases: Pareidolia relates to Apophenia, the tendency to perceive connections between unrelated things, and Illusory Correlation, where we wrongly assume a relationship between two unrelated events.

Countermeasures: A healthy dose of skepticism can counter pareidolia. Question your perceptions, especially when you find significance in randomness. Scientific methodology and evidence-based reasoning can also help discern fact from fantastical fiction.

Key Takeaways: While pareidolia reflects our brain's impressive pattern-recognition capabilities, it can lead us to see faces in clouds or hear messages in white noise. Staying skeptical and grounded in evidence can help us distinguish meaningful patterns from mere illusions.

References:

Liu, J., Li, J., Feng, L., Li, L., Tian, J., & Lee, K. (2014). Seeing Jesus in toast: Neural and behavioral correlates of face pareidolia. Cortex, 53, 60-77. https://doi.org/10.1016/j.cortex.2014.01.013

Shermer, M. (2008). Patternicity: Finding meaningful patterns in meaningless noise. Scientific American, 299(5), 48. https://doi.org/10.1038/scientificamerican1108-48

Peak-End Rule

Definition: The Peak-End Rule is the cognitive bias where people judge an experience largely based on how they felt at its peak and at its end, rather than the total sum or average of every moment of the experience. In simpler terms, we remember the best of times, the worst of times, and the last of times – everything else gets a bit fuzzy.

Description: This psychological heuristic suggests that our perception of past events doesn't take into account the duration. We focus on the most intense points and the end of the experience when evaluating it. It's like remembering a roller coaster ride for its scariest drop and the sigh of relief at the end, but forgetting about the rest of the loop-de-loops.

Examples: In the realm of religion, a single profound religious experience or the last one before the recollection could have a disproportionate impact on a person's religious belief, overshadowing years of less intense experiences. Politically, a candidate's one outstanding success, major failure, or recent actions could dominate voters' perceptions. In social issues, an activist's most notable achievement or recent actions could significantly influence public opinion about their entire career.

Impact: The Peak-End Rule can lead to skewed perceptions and judgments. It can result in overemphasis on a few intense or recent moments, neglecting the overall experience. It's why a speech may be remembered for its climactic moment and powerful conclusion, even if the rest was less than stellar.

Related Biases: The Peak-End Rule is related to the Recency Effect, where the most recent information is remembered best, and the Availability Heuristic, where the most readily available experiences influence decision-making.

Countermeasures: One way to combat the Peak-End Rule is to consciously consider the entire duration and breadth of an experience, not just its peaks and end. Try to evaluate events in their entirety rather than letting the most dramatic or recent moments dominate your perception.

Key Takeaways: The Peak-End Rule suggests we evaluate experiences not on their entirety but on their peak moments and how they end. It's essential to be aware of this bias when recalling past events or making decisions based on past experiences.

References:

Kahneman, D., Fredrickson, B. L., Schreiber, C. A., & Redelmeier, D. A. (1993). When More Pain Is Preferred to Less: Adding a Better End. Psychological Science, 4(6), 401–405. https://doi.org/10.1111/j.1467-9280.1993.tb00589.x

Redelmeier, D. A., & Kahneman, D. (1996). Patients' memories of painful medical treatments: Real-time and retrospective evaluations of two minimally invasive procedures. Pain, 66(1), 3-8. https://doi.org/10.1016/0304-3959(96)02994-6

Perceptual Bias

Definition: Perceptual bias is a psychological tendency to interpret and understand information based on one's existing beliefs, experiences, and expectations, often resulting in a distorted view of reality. In essence, it's like donning tinted glasses that color everything you see.

Description: This cognitive bias involves allowing our preconceived notions and personal interpretations to shape our perception of reality. It essentially acts as a cognitive filter, allowing us to see what we want or expect to see, while simultaneously blinding us to differing viewpoints or facts.

Examples: On the religious front, a believer might perceive benign events as divine intervention because of their faith, while a nonbeliever might interpret the same events as mere coincidence. In politics, a conservative might perceive progressive policies as harmful and impractical, while a liberal might see them as beneficial and necessary. In social issues, a person might perceive a protest as either a commendable expression of free speech or a disruptive nuisance, depending on their personal beliefs.

Impact: Perceptual bias can profoundly influence our decisions and judgments. It can lead to misunderstanding, misinterpretation, and at times, can result in unfair treatment or discrimination. Essentially, it might just turn you into the embodiment of that old adage "judge a book by its cover."

Related Biases: Confirmation bias, where we favor information that confirms our preexisting beliefs, and selective perception, where we selectively interpret what we see based on our interests, background, and attitudes, are closely related to perceptual bias.

Countermeasures: Awareness is the first step. Recognizing the existence of perceptual bias is critical in combating its influence. From there, one can make an effort to expose oneself to diverse viewpoints and consciously seek out information that challenges existing beliefs.

Key Takeaways: Perceptual bias can seriously skew our perception of reality. It's crucial to be aware of this bias and make conscious efforts to consider multiple viewpoints and not let our preconceived notions color our understanding of the world around us.

References:

Kruger, J., & Gilovich, T. (2004). Actions, intentions, and self-assessment: The road to self-enhancement is paved with good intentions. Personality and Social Psychology Bulletin, 30(3), 328–339. https://doi.org/10.1177/0146167203259932

Pronin, E., Lin, D. Y., & Ross, L. (2002). The Bias Blind Spot: Perceptions of Bias in Self Versus Others. Personality and Social Psychology Bulletin, 28(3), 369–381. https://doi.org/10.1177/0146167202286008

Picture Superiority Effect

Definition: The picture superiority effect is a cognitive bias where individuals tend to better remember images than words. This is essentially the "a picture is worth a thousand words" effect in cognitive psychology.

Description: Our brains, being the wonderfully efficient yet sometimes lazy organs they are, process visual information more easily than text. The picture superiority effect illustrates how images are more easily remembered and recalled than mere words. This bias is related to the dual-coding theory, suggesting that images are coded twice in our memory – once as an image and again as a semantic (word-based) code, making them more memorable.

Examples: In politics, remember those iconic campaign posters? They work so well because of the picture superiority effect. Likewise, in religious texts, illustrated parables may be easier to remember than written ones. Social media platforms, like Instagram, have skyrocketed in popularity partly due to this effect. Our brains are simply smitten with the pictures.

Impact: The picture superiority effect can have significant impacts on how information is conveyed, marketed, and remembered. It influences decision-making in various contexts, from which product to buy, which political candidate to vote for, to how effectively we can recall safety procedures.

Related Biases: The Von Restorff effect (or isolation effect), which predicts that when multiple similar objects are present, the one that differs from the rest is most likely to be remembered, is somewhat related. The picture superiority effect often makes the image the standout.

Countermeasures: To counteract this bias, especially in situations where text-based information is crucial, it's important to ensure key points are emphasized and reiterated, possibly with the use of complementary visuals.

Key Takeaways: We're visually wired creatures, making the picture superiority effect a powerful tool in communication, marketing, and education. However, it's important to remember that key textual information should be emphasized to ensure it doesn't become the forgettable wallflower at the visual party.

References:

Paivio, A. (1971). Imagery and Verbal Processes. Holt, Rinehart, and Winston.

Shepard, R. N. (1967). Recognition memory for words, sentences, and pictures. Journal of Verbal Learning and Verbal Behavior, 6(1), 156-163. doi:10.1016/S0022-5371(67)80067-7

Waddill, P. J., & McDaniel, M. A. (1992). Pictorial Enhancement of Recall: Limitations Imposed by Picture Type and Comprehension Skill. Memory & Cognition, 20(5), 472–482. doi: 10.3758/bf03210936

Planning Fallacy

Definition: The planning fallacy is a cognitive bias in which individuals underestimate the time it will take to complete a future task, despite past experience showing that similar tasks generally take longer than planned.

Description: This delightful optimism we seem to have about our future productivity can be quite the charmer. Despite the voice of experience whispering in our ears, "You know it took you twice as long last time, right?", we confidently assure ourselves that this time it'll be different. This is the heart of the planning fallacy – underestimating the time or resources required to complete a task.

Examples: For an example, let's travel to the realm of politics where election promises are as plentiful as the grains of sand on a beach. Politicians routinely fall prey to the planning fallacy, promising grand projects will be completed within a certain timeframe, only to face delays and budget overruns. Sound familiar?

In religious contexts, imagine a congregation planning a community outreach event. Despite last year's event taking weeks longer to plan than they anticipated, they're certain they can get it done in half the time this year.

Impact: The impact of the planning fallacy can range from mildly frustrating (an overdue library book) to disastrous (multibillion-dollar overruns on public projects). It impacts our personal lives, professional environments, and larger societal structures, often leading to missed deadlines, budget overflows, and stress.

Related Biases: Optimism bias is a close cousin, which refers to our tendency to believe that our future will be better and we're less at risk of experiencing negative events compared to others. Both are partly to blame for our overly rosy planning forecasts.

Countermeasures: Counteracting this bias involves adjusting time and resource estimates based on past experience and allowing for contingencies. Involving others in the planning process can also help, as they may see potential obstacles or delays you may have overlooked.

Key Takeaways: The planning fallacy is a common misstep, making us overly optimistic about our future productivity. Being aware of this bias, learning from past experiences, and making room for the unexpected can help keep our plans and projects on track.

References:

Buehler, R., Griffin, D., & Ross, M. (1994). Exploring the "planning fallacy": Why people underestimate their task completion times. Journal of Personality and Social Psychology, 67(3), 366–381. doi:10.1037/0022-3514.67.3.366

Kahneman, D., & Tversky, A. (1977). Intuitive prediction: Biases and corrective procedures. In DECISIONS AND DESIGNS INC MCLEAN VA (Ed.), Technical Report. doi:10.21236/ada047747

Positivity Effect

Definition: The positivity effect is a cognitive bias that leads individuals, particularly older adults, to favor positive over negative information in their perceptions, memory recall, and decision-making processes.

Description: It seems that as we age, we grow increasingly fond of rose-tinted spectacles. Our mental scrapbooks become filled with sunny memories, while the gloomy ones somehow get misplaced. This is the positivity effect – a natural tendency to tilt towards the positive side of life, especially as we accumulate more birthdays.

Examples: Let's take a quick detour to our society's favorite pastime - politics. Ever noticed how seasoned politicians effortlessly recall their victories while the not-so-glorious moments seem to have taken a long sabbatical? That's the positivity effect in action.

In the context of religion, long-standing members of a congregation might have a stronger recall of positive community activities, while memories of internal conflicts or disagreements conveniently fade into the background.

Impact: While the positivity effect might seem like a jovial, harmless old chap, it can impact our decision-making processes and perception of reality. Overemphasis on positive experiences might lead to overlooking risks or neglecting critical but negative information. On the flip side, it could contribute to greater life satisfaction and wellbeing in older age.

Related Biases: It's closely linked to the Pollyanna principle, which proposes that people (not just the older folks) naturally lean towards positivity when thinking about the past.

Countermeasures: Being aware of the positivity bias can encourage a more balanced perspective. Aiming for realistic optimism, which involves acknowledging both positive and negative aspects, can help in maintaining a fair view of situations.

Key Takeaways: The positivity effect illustrates our inherent lean towards the sunny side of life, especially as we age. While it can enhance our life satisfaction, maintaining a balanced perspective is essential to avoid overlooking critical aspects of reality.

References:

Charles, S. T., Mather, M., & Carstensen, L. L. (2003). Aging and emotional memory: The forgettable nature of negative images for older adults. Journal of Experimental Psychology: General, 132(2), 310–324. doi:10.1037/0096-3445.132.2.310

Reed, A. E., & Carstensen, L. L. (2012). The Theory Behind the Age-Related Positivity Effect. Frontiers in psychology, 3, 339. doi:10.3389/fpsyg.2012.00339

Post Hoc Ergo Propter Hoc Fallacy

Definition: The post hoc ergo propter hoc fallacy (or, for those of us who didn't take Latin in high school, "after this, therefore because of this") is a logical misstep where one concludes that because one event followed another, the first event caused the second.

Description: When things happen in quick succession, it's tempting to link them together like beads on a string. But life isn't a simple connect-the-dots puzzle. Sometimes, two dots just happen to fall in line without having a line between them. This is the essence of the post hoc ergo propter hoc fallacy - assuming a causal link where none exists, solely based on the order of events.

Examples: Let's dive into some examples, shall we? In the world of politics, a newly elected leader might claim credit for an economic boom, even though the policies responsible were enacted by their predecessor.

In the realm of religion, a parishioner might attribute their unexpected recovery from an illness to the prayer circle that took place the day before.

As for social issues, imagine someone asserting that a rise in video game popularity leads to increased violence, simply because the two trends coincided.

Impact: The post hoc fallacy can lead to misguided beliefs and poorly informed decisions. We may place unwarranted faith in ineffective practices or overlook the real causal factors in a situation. Misunderstanding causation can hinder our ability to predict future events and make effective changes.

Related Biases: This fallacy is a member of the larger family of causal fallacies, which also includes the cum hoc ergo propter hoc fallacy ("with this, therefore because of this").

Countermeasures: Rigorous critical thinking can help us avoid this pitfall. When evaluating potential causes and effects, we need to look for evidence beyond mere temporal sequence. Correlation doesn't imply causation, and neither does sequence!

Key Takeaways: The post hoc ergo propter hoc fallacy is a misstep in causal reasoning, assuming that because one event follows another, the first must have caused the second. It's a trap that can lead to misguided beliefs and decisions, so stay vigilant and remember: sequence is not the same as causation!

References:

Shaffer, J. (2005). The Post Hoc Ergo Propter Hoc Fallacy. Informal Logic, 25(3). doi:10.22329/il.v25i3.2289

Chen, E. E., & Wojcik, S. P. (2016). A practical guide to big data research in psychology. Psychological Methods, 21(4), 458–474. doi:10.1037/met0000111

Post-Purchase Rationalization

Definition: Post-Purchase Rationalization is a cognitive bias where individuals justify their purchases, particularly expensive or regrettable ones, by convincing themselves of the inherent value or benefits that weren't initially apparent or factual.

Description: Essentially, it's a mental gymnastics routine, performed to maintain a positive self-image and alleviate cognitive dissonance. You might say it's the "sour grapes" phenomenon, but in reverse and with price tags attached.

Examples: In politics, for instance, let's say a candidate secures a high-priced campaign consultant. After an unsuccessful election, they might insist that the consultant's advice was invaluable for future campaigns.

In religion, consider a parishioner who donates a sizable sum to their place of worship and later convinces themselves that their generosity will ensure a cushy afterlife.

On a social scale, imagine a homeowner justifying their overpriced, environmentally-friendly vehicle purchase by emphasizing its potential impact on decreasing their carbon footprint, despite it straining their budget.

Impact: This bias can result in unnecessary spending, bad investment decisions, and overall financial mismanagement. It also promotes the continuation of harmful patterns of decision-making, as individuals may fail to learn from their financial mistakes.

Related Biases: Post-Purchase Rationalization is closely related to the Sunk Cost Fallacy, where one is reluctant to abandon an endeavor because they've invested heavily in it, regardless of future prospects.

Countermeasures: To counter this bias, practice mindful spending: consider your purchases carefully and try to base them on needs rather than impulsive wants. A period of "cooling off" before making significant purchases can also be beneficial.

Key Takeaways: Post-Purchase Rationalization is the mental gymnastics routine we perform after splurging on a purchase we might regret. To avoid financial discomfort and promote more rational decision-making, consider practicing mindful spending and giving yourself a "cooling-off" period before big purchases.

References:

Buehler, R., Griffin, D., & Ross, M. (1994). Exploring the "planning fallacy": Why people underestimate their task completion times. Journal of Personality and Social Psychology, 67(3), 366–381. doi:10.1037/0022-3514.67.3.366

Arkes, H. R., & Blumer, C. (1985). The psychology of sunk cost. Organizational Behavior and Human Decision Processes, 35(1), 124–140. doi:10.1016/0749-5978(85)90049-4

Pratfall Effect

Definition: The Pratfall Effect is a cognitive bias in which people's attractiveness or likability is increased or decreased based on how they handle making mistakes. If a person is perceived as competent, a minor goof can make them seem more appealing, more human, and less intimidating.

Description: The human mind loves a good paradox. We adore perfection, but we're also charmed by the occasional stumble. The Pratfall Effect, named after the pratfall comedy trope, is evidence of this. It suggests that when competent individuals make a minor blunder, they're often seen as more appealing rather than less.

Examples: In politics, a well-regarded leader making an innocent, non-critical mistake (like mispronouncing a word in a speech) may seem more relatable and human to voters.

In religion, a clergy member admitting to a benign personal flaw could foster a deeper connection with their congregation.

And socially, think about how endearing it can be when an otherwise flawless friend spills coffee on themselves, reinforcing the fact that, yes, they are human.

Impact: This bias influences how we perceive, judge, and interact with others, affecting social dynamics and relationships. It can, however, lead to excessive self-deprecation or feigned incompetence, as people might purposely make mistakes to appear more likable.

Related Biases: This effect is closely linked with the Halo Effect, where our impression of someone in one area (like physical attractiveness) influences our judgment of their other traits.

Countermeasures: To mitigate this bias, strive for self-awareness and authenticity. It's okay to admit mistakes, but don't force them for the sake of likability. Also, when evaluating others, try to separate isolated instances from overall competence or character.

Key Takeaways: The Pratfall Effect reminds us that perfection can be a bit off-putting. A small blunder from a competent individual can make them seem more human and likable. But remember: authenticity is key.

References:

Aronson, E., Willerman, B., & Floyd, J. (1966). The effect of a pratfall on increasing interpersonal attractiveness. Psychonomic Science, 4(6), 227–228. doi:10.3758/BF03342263

Nisbett, R. E., & Wilson, T. D. (1977). The halo effect: Evidence for unconscious alteration of judgments. Journal of Personality and Social Psychology, 35(4), 250–256. doi:10.1037/0022-3514.35.4.250

Primacy Effect

Definition: The Primacy Effect is a cognitive bias that entails the tendency to remember and give more importance to the first items or information we encounter, compared to those we encounter later.

Description: This intriguing bias underscores our brain's knack for the theatrical. It seems the opening act leaves a stronger impression than the rest of the show. The Primacy Effect suggests that we are more likely to recall and be influenced by the initial information we receive, whether it's the first item on a list or the first impression of a person.

Examples: In politics, voters often recall first impressions of candidates more vividly than subsequent events or speeches.

In social issues, initial news reports can color our perceptions of ongoing events, even if later reports provide more accurate information.

And in religion, the first doctrines or beliefs introduced to someone often set the stage for their future understanding and interpretation of that religion.

Impact: This bias can lead to skewed judgments and decisions, as later but possibly more important or accurate information may be undervalued.

Related Biases: The Primacy Effect is closely related to the Recency Effect, where the most recently presented items or experiences are the ones most remembered.

Countermeasures: Being aware of this bias can help to counteract it. Make a conscious effort to give equal weight to all information received, not just the first. Review all available information before making decisions or forming impressions.

Key Takeaways: The Primacy Effect reminds us that our brains tend to put a spotlight on the first piece of information they receive. However, the rest of the story is just as important, and we should strive to consider all information equally.

References:

Murdock, B. B. (1962). The serial position effect of free recall. Journal of Experimental Psychology, 64(5), 482–488. doi:10.1037/h0045106

Luchins, A. S. (1957). Primacy-recency in impression formation. In C. I. Hovland (Ed.), The order of presentation in persuasion (pp. 33-61). New Haven, CT: Yale University Press.

Priming

Definition: Priming is a cognitive bias where exposure to one stimulus influences a response to a subsequent stimulus, without conscious guidance or intention.

Description: Think of your mind as an overeager association game player, making connections between stimuli faster than a squirrel on espresso. That's priming. It's when exposure to one thing (like a word, image, or event) influences how you respond to another thing. The fun twist? It all happens without your conscious awareness.

Examples: In politics, a candidate's image on a billboard (prime) might subconsciously influence voters' opinions when they later see the candidate in a debate (response).

In social issues, media stories (prime) can impact public perceptions and responses to certain groups or events.

And in religion, a sermon or scripture reading (prime) can influence a believer's later thoughts or actions (response).

Impact: Priming can shape our perceptions, attitudes, and behavior in subtle but potentially significant ways. It can sway public opinion, influence consumer behavior, and affect interpersonal interactions.

Related Biases: The Mere Exposure Effect (a preference for familiar things) and Framing Effect (drawing different conclusions based on how the same information is presented) are closely related biases.

Countermeasures: To mitigate the influence of priming, strive for awareness and conscious decision-making.

Question why you hold certain attitudes or make specific choices and seek out diverse sources of information to avoid being overly influenced by a single perspective.

Key Takeaways: Priming reminds us that our brains can be nudged in certain directions by prior stimuli, shaping our thoughts and actions in ways we might not realize. By staying mindful and questioning our responses, we can aim to make more informed and balanced decisions.

References:

Bargh, J. A., & Pietromonaco, P. (1982). Automatic information processing and social perception: The influence of trait information presented outside of conscious awareness on impression formation. Journal of Personality and Social Psychology, 43(3), 437–449. doi:10.1037/0022-3514.43.3.437

Tulving, E., & Schacter, D. L. (1990). Priming and human memory systems. Science, 247(4940), 301–306. doi:10.1126/science.2296719

Pro-Innovation Bias

Definition: Pro-innovation bias is the inclination to view new ideas or innovations as superior to existing concepts or methods, simply because they're new.

Description: Imagine that you're at a tech expo, and someone presents a shiny new gadget, proclaiming it will change everything. Your eyes light up, and you're instantly captivated. That's pro-innovation bias in action. This bias tends to nudge us into overvaluing the benefits of newness and innovation, often while overlooking their limitations or potential downsides.

Examples: In politics, it could be the rush to implement a novel policy or political structure without fully considering its consequences.

In social issues, it could be the tendency to enthusiastically embrace new social trends or movements without critically examining their implications.

In religion, it might manifest as a propensity to favor new interpretations or practices over traditional ones, simply because they bring a fresh perspective.

Impact: While innovation is undoubtedly vital, the pro-innovation bias can lead to hasty decisions, oversight of potential issues, and a disregard for the merits of existing methods or ideas.

Related Biases: This bias is akin to the novelty effect, where people show an increased level of response to anything new or unfamiliar, and neophilia, the love of anything new or novel.

Countermeasures: Encourage critical thinking and thorough examination of all aspects of an innovation. Don't discount something merely because it's old or embrace something only because it's new.

Key Takeaways: Innovation is crucial, but not all that glitters is gold. Pro-innovation bias can blind us to potential pitfalls, so always critically evaluate the new and shiny before jumping on board.

References:

Rogers, E. M. (2003). Diffusion of innovations (5th ed.). New York: Free Press.

Fichman, R. G., & Kemerer, C. F. (1999). The illusory diffusion of innovation: An examination of assimilation gaps. Information Systems Research, 10(3), 255-275. doi:10.1287/isre.10.3.255

Projection Bias

Definition: Projection bias is a cognitive bias that leads individuals to assume others share the same beliefs, attitudes, and values they do, even when evidence suggests otherwise.

Description: Put simply, it's a case of thinking, "Hey, I like sardines on pizza, who wouldn't?" (spoiler: plenty of people wouldn't). It's a form of egocentric bias, where we use our own feelings and experiences as a model for others, often leading to misunderstanding and miscommunication.

Examples: In politics, projection bias might be thinking that everyone in your party supports a particular candidate or policy just because you do.

In religion, it can manifest as assuming that others share your beliefs about morality or spiritual matters.

When it comes to social issues, this bias can lead to presumptions that everyone in a particular group must share the same opinions and attitudes.

Impact: Projection bias can hinder productive dialogue and create discord, leading to polarized societies and "us-versus-them" mentality. It can distort our understanding of others' views, leading to flawed decision-making and inaccurate predictions of others' behaviors.

Related Biases: False-consensus effect (overestimating the extent to which others share our beliefs) and egocentric bias (over-reliance on one's own perspective) are closely related to projection bias.

Countermeasures: Enhance self-awareness, encourage perspective-taking, promote empathy, and foster diverse social interactions to help counter this bias.

Key Takeaways: Your favorite flavor of ice cream might not be everyone's cup of tea. Just as with frozen desserts, so it is with beliefs and attitudes - variety is the spice of life.

References:

Ross, L., Greene, D., & House, P. (1977). The false consensus effect: An egocentric bias in social perception and attribution processes. Journal of Experimental Social Psychology, 13(3), 279–301. https://doi.org/10.1016/0022-1031(77)90049-X

Krueger, J., & Clement, R. W. (1994). The truly false consensus effect: An ineradicable and egocentric bias in social perception. Journal of Personality and Social Psychology, 67(4), 596–610. https://doi.org/10.1037/0022-3514.67.4.596

Proportionality Bias

Definition: Proportionality bias is a cognitive tendency to assume big events must have big causes. That's right, folks, we're not talking about oversizing your morning coffee here; this is about over-sizing causal relationships.

Description: Imagine thinking every puddle is the result of a storm, not just some kid spilling their water bottle. That's proportionality bias. It's one of the reasons why conspiracy theories are so popular, because they often propose grandiose causes for significant events, rather than more mundane or complex explanations.

Examples: In the world of religion, people often attribute significant life events, like the birth of a child or a near-death experience, to divine intervention.

In politics, you might think that a major policy failure was due to a grand, malevolent conspiracy, rather than a series of minor bureaucratic blunders.

For social issues, proportionality bias could lead someone to attribute systemic problems, like crime rates or unemployment, to a single factor, such as moral decay or laziness, rather than acknowledging the intricate web of contributing factors.

Impact: Proportionality bias can oversimplify complex issues and hinder productive problem-solving. It promotes binary thinking and can contribute to polarization and division.

Related Biases: The conjunction fallacy (the belief that specific conditions are more probable than a single general one) and illusory correlation (perceiving a relationship between variables even when no such relationship exists) are related biases.

Countermeasures: Awareness of this bias is key, along with a willingness to engage with complex, multi-causal explanations for significant events. It's not as sexy as a conspiracy theory, but it's more likely to get you closer to the truth.

Key Takeaways: Big effects often have small, complex causes, not just one grand cause. So, put away the tinfoil hat and embrace the complex reality.

References:

Leman, P. J., & Cinnirella, M. (2007). A major event has a major cause: Evidence for the role of heuristics in reasoning about conspiracy theories. Social Psychological Review, 9(2), 18–28.

Brotherton, R., & French, C. C. (2014). Belief in conspiracy theories and susceptibility to the conjunction fallacy. Applied Cognitive Psychology, 28(2), 238–248. https://doi.org/10.1002/acp.2995

Prototype Heuristic

Definition: Prototype heuristic is a cognitive shortcut where we judge by categorizing items or people based on how similar they are to a prototype, or standard mental representation, in our mind. It's kind of like how your uncle at the family reunion insists that your cousin is just a "younger version" of you. No, Uncle Jerry, we just both have brown hair.

Description: This cognitive shortcut aids in rapid decision-making but can also result in errors, oversimplifications, or stereotyping. It's a natural, if occasionally unhelpful, part of how our brains make sense of the world.

Examples: In religion, people may perceive others as "religious" or "not religious" based on how closely they adhere to the prototypical behaviors of a religious person.

In politics, voters often categorize politicians into "conservative" or "liberal" camps based on a few key policy positions, without acknowledging the complexity of political belief systems.

In social issues, prototype heuristic could lead someone to stereotype individuals based on their membership in a particular group, leading to assumptions about their characteristics, beliefs, or behaviors.

Impact: Prototype heuristic can facilitate snap judgements and stereotyping, obscuring the individual characteristics and complexities that defy categorization. It can contribute to societal divisions and misunderstandings.

Related Biases: Stereotyping, out-group homogeneity bias (seeing members of an out-group as more similar to each other than they really are), and the representativeness heuristic (judging the probability of an event based on how similar it is to your mental prototype) are related biases.

Countermeasures: Increasing awareness of this bias and striving to see individuals and situations in their own unique context can help to reduce the influence of the prototype heuristic. Just because your cousin has brown hair doesn't mean she shares your fervor for antique stamp collecting.

Key Takeaways: Recognize that prototypes are shortcuts, not destinations. Strive to see the individual, not just the category they might fit into.

References:

Smith, E. R., & Zarate, M. A. (1992). Exemplar-based model of social judgment. Psychological Review, 99(1), 3–21. https://doi.org/10.1037/0033-295X.99.1.3

Sherif, M., & Hovland, C. I. (1961). Social judgment: Assimilation and contrast effects in communication and attitude change. Yale University Press

Pseudocertainty Effect

Definition: The pseudocertainty effect is a cognitive bias in which individuals perceive an outcome as certain while it is, in fact, uncertain. It's like thinking you're a shoo-in for the promotion because you've been with the company for ten years when, in reality, the boss's pet goldfish has a better chance.

Description: This bias emerges from our tendency to simplify complex situations by creating an illusion of certainty. It usually happens in multi-stage decisions where, if one stage is perceived as certain, we overlook the uncertainty of subsequent stages. It's kind of like the mental equivalent of putting your fingers in your ears and humming loudly.

Examples: In religion, one might believe that by performing certain rituals or practices, they're guaranteed a pleasant afterlife. In politics, voters might be overconfident about their candidate's victory because they've won the primary, disregarding the uncertainties of the general election. On social issues, activists might fall into the trap of assuming victory once a law is passed, neglecting the challenges of implementation and societal change.

Impact: The pseudocertainty effect can lead to overconfidence, poor risk assessment, and inadequate preparation for outcomes that deviate from our expectations. Your metaphorical bubble could be popped rather abruptly.

Related Biases: Overconfidence bias, illusion of control, and optimism bias are all cousins of pseudocertainty. They all enjoy pretending things are going splendidly when they may not be.

Countermeasures: Always question your assumptions. Embrace uncertainty and consider a range of potential outcomes, not just the most favorable ones. And remember, the boss's goldfish is always watching.

Key Takeaways: Certainty is comforting, but illusions of certainty can lead us astray. Question your assumptions, plan for a range of outcomes, and don't underestimate the boss's goldfish.

References:

Tversky, A., & Kahneman, D. (1981). The framing of decisions and the psychology of choice. Science, 211(4481), 453–458. https://doi.org/10.1126/science.7455683

Weber, E. U., & Johnson, E. J. (2009). Mindful judgment and decision making. Annual Review of Psychology, 60, 53–85. https://doi.org/10.1146/annurev.psych.60.110707.163633

Pseudo-opinion

Definition: A pseudo-opinion is an alleged viewpoint that is expressed without substantial knowledge or understanding of the subject matter. It's like telling someone how to raise their children when your only child-rearing experience is with a houseplant.

Description: Pseudo-opinions are often formed under societal pressure to have an opinion on everything. This is fueled by our need to fit in and our desire to avoid appearing uninformed. It's the cognitive equivalent of bluffing your way through a high school English essay about a book you never read.

Examples: For religion, a person might debate the interpretation of a religious text they've never read. In politics, an individual might passionately discuss policy issues based on headlines alone, without understanding the complexities involved. In social issues, one might assert strong views about a controversial topic they've only heard about in passing.

Impact: Pseudo-opinions can muddy public discourse, spread misinformation, and hinder genuine understanding. They can polarize societies and generate unfounded hostility. Also, if you're found out, you could end up looking like a bit of a nincompoop.

Related Biases: The Dunning-Kruger effect, where individuals with low ability at a task overestimate their ability, is like a close, but equally clueless, relative of the pseudo-opinion.

Countermeasures: Foster a culture where it's okay to say "I don't know," and encourage informed discussion over opinionated chatter. Spend time researching before forming an opinion. Your houseplant will thank you for it.

Key Takeaways: It's better to be thought a fool than to open your mouth and remove all doubt. Think before you speak, research before you opine, and take good care of your houseplant.

References:

Fischer, P., Schulz-Hardt, S., & Frey, D. (2008). Selective exposure and decision framing: The impact of gain and loss framing on confirmatory information search after decisions. Journal of Experimental Social Psychology, 44(2), 312-320. https://doi.org/10.1016/j.jesp.2007.06.001

Schuman, H., & Presser, S. (1981). Questions and Answers in Attitude Surveys: Experiments on Question Form, Wording, and Context. Academic Press.

Psychological Reactance

Definition: Psychological reactance is the emotional response triggered when individuals perceive that their freedoms or choices are being restricted. Think of it as the mental equivalent of a two-year-old's tantrum when told they can't have another cookie.

Description: This phenomenon occurs when people perceive a threat to their autonomy, leading to a motivational state that seeks to restore the threatened freedom. It's kind of like being told you can't do something and suddenly wanting to do it twice as much, just to prove a point.

Examples: In religion, some may rebel against dogmatic practices, seeing them as a violation of personal freedom. In politics, people might resist policies seen as overreaching or restrictive, like mandatory mask-wearing during a pandemic. In social issues, teenagers may do the exact opposite of what their parents advise, purely out of reactance.

Impact: Psychological reactance can influence decision-making and behavior, often leading to counterproductive results. Like throwing a tantrum in the middle of the supermarket because you can't have a cookie. Not a good look for anyone involved.

Related Biases: The boomerang effect, where a person does the opposite of what they're advised or persuaded to do, is a close cousin of psychological reactance.

Countermeasures: Promote choices and alternatives rather than imposing strict rules. Framing is key - instead of saying "You must," try "You could." It's a Jedi mind trick for cognitive bias.

Key Takeaways: People don't like being told what to do. Recognize this, promote freedom of choice, and avoid unnecessary restrictions. And remember, sometimes a cookie is just a cookie.

References:

Brehm, J. W. (1966). A theory of psychological reactance. Academic Press.

Miron, A. M., & Brehm, J. W. (2006). Reactance Theory - 40 Years Later. Zeitschrift für Sozialpsychologie, 37(1), 9–18. https://doi.org/10.1024/0044-3514.37.1.9

Pygmalion Effect

Definition: The Pygmalion Effect is a psychological phenomenon wherein high expectations lead to improved performance in a given area. In other words, if you think you're a rockstar, you just might become one.

Description: This bias is about self-fulfilling prophecies, where our beliefs influence our behavior in a way that makes the beliefs come true. The term originates from a Greek myth where Pygmalion, a sculptor, fell in love with a statue he created, and it came to life. Yes, reality is stranger than fiction... or is that the other way around?

Examples: In religion, a charismatic leader's belief in their followers' righteousness could inspire them to live more virtuously. In politics, public faith in a candidate's capability can drive them to exceed their own expectations. In social issues, a teacher's belief in a student's potential can lead to improved academic performance. It's the power of belief, folks!

Impact: The Pygmalion Effect can have profound impacts on performance, motivation, and self-esteem. However, it can also create pressure to live up to unrealistic expectations, which might lead to a spectacular face-plant.

Related Biases: The Golem Effect, where low expectations lead to decreased performance, is the pessimistic sibling of the Pygmalion Effect.

Countermeasures: Encourage high expectations, but keep them realistic. Create a positive and supportive environment that nurtures belief and self-confidence without setting people up for failure.

Key Takeaways: Beliefs can be powerful drivers of action. Manage expectations wisely to harness the positive aspects of the Pygmalion Effect, and beware of setting unrealistic standards. Remember, Rome wasn't built in a day, but you can certainly start laying the bricks.

References:

Rosenthal, R., & Jacobson, L. (1968). Pygmalion in the Classroom. The Urban Review, 3, 16–20. https://doi.org/10.1007/BF02322211

McNatt, D. B. (2000). Ancient Pygmalion joins contemporary management: A meta-analysis of the result. Journal of Applied Psychology, 85(2), 314–322. https://doi.org/10.1037/0021-9010.85.2.314

Reactive Devaluation

Definition: Reactive devaluation is a cognitive bias where an individual assigns less value to an offer simply because it originates from an adversary or someone they dislike. In essence, it's the principle of 'if they suggested it, it must be a bad idea'.

Description: This bias occurs because of the influence of personal feelings on objective judgement. If we have a preconceived negative view of a person or group, we instinctively devalue their proposals, even if they're objectively good or in our best interest.

Examples: In politics, a brilliant bill could be dismissed purely because it was proposed by the opposition party. In religion, a worthy suggestion from a rival faith might be devalued or ignored. And socially, how often do we dismiss sound advice from an annoying co-worker just because, well, they're annoying?

Impact: Reactive devaluation can hinder negotiations, obstruct progress, and contribute to a hostile environment. It's the equivalent of cutting off your nose to spite your face.

Related Biases: This bias is akin to the 'Not Invented Here' bias, where people devalue ideas that originated elsewhere.

Countermeasures: Recognizing the bias is the first step. Try to evaluate proposals based on their merit and not their source. It's tough, especially when you're convinced the source couldn't find their way out of a wet paper bag. But do try.

Key Takeaways: Don't let personal feelings cloud your judgement. Evaluate suggestions based on their value, not their origin. It's like judging a book by its content, not by the color of its cover. You might be surprised.

References:

Ross, L., & Stillinger, C. (1991). Barriers to conflict resolution. Negotiation journal, 7(4), 389-404. https://doi.org/10.1111/j.1571-9979.1991.tb00634.x

Stein, J. G. (1989). Getting to the table. The processes of international prenegotiation. US Institute of Peace Press.

Recency Effect

Definition: The recency effect is a cognitive bias in which recent information is more likely to be remembered and thus considered more significant than older information. In essence, it's like a mental "What have you done for me lately?"

Description: This bias is part of the broader 'serial position effect' in psychology, which states that the position of an item in a sequence significantly affects recall. The recency effect makes us more likely to recall items at the end of the list. It's kind of like our brains are a leaky bucket – we're better at holding onto the last things put in there.

Examples: In politics, an incumbent might gain an advantage by making a popular decision close to an election. In social issues, a recent scandal can overshadow years of good work. And in religion, a recent miraculous event could stir more faith than ancient texts.

Impact: The recency effect can lead to short-term thinking, where recent events disproportionately influence our decisions and judgments. Long-term planning can become a thing of myth and legend. Remember the tortoise and the hare? Yeah, neither do we.

Related Biases: It is closely related to the primacy effect, where the first information is remembered more clearly.

Countermeasures: One way to counteract this bias is by actively recalling older information and considering the big picture before making decisions. A technique? Try keeping a journal or use a good old spreadsheet. Just don't forget where you put it.

Key Takeaways: Remember to factor in the big picture and not just recent events when making decisions. And try to remember the tortoise and the hare; it's a good story.

References:

Murdock, B. B. (1962). The serial position effect of free recall. Journal of experimental psychology, 64(5), 482. https://doi.org/10.1037/h0045106

Murre, J. M., & Dros, J. (2015). Replication and analysis of Ebbinghaus' forgetting curve. PloS one, 10(7), e0120644. https://doi.org/10.1371/journal.pone.0120644

Reification (Concretism, Hypostatization, or the Fallacy of Misplaced Concreteness)

Definition: Reification, also known as the Fallacy of Misplaced Concreteness, is a cognitive bias where we treat abstract concepts as if they were concrete or tangible entities. Essentially, it's when we take our metaphors too literally and start shopping for a ticket to "Cloud Nine".

Description: Reification often occurs when a word or phrase that originally referred to an idea or conceptual process comes to be used as if it refers to a physical entity. It's like when you catch yourself checking your pockets for your train of thought.

Examples: In politics, concepts like "the economy" or "the will of the people" are often reified, talked about as if they were single, tangible things rather than complex, multi-faceted systems or diverse groups of individuals. In social discourse, "society" is often blamed or praised as if it's an individual actor with intentions. In religion, abstract ideas such as 'sin' or 'grace' may be discussed as if they were physical objects.

Impact: Reification can lead to oversimplified thinking, faulty reasoning, and misunderstandings. You wouldn't want to build a house on a metaphorical foundation, now would you?

Related Biases: It's closely related to the fallacy of reification, which is the error of treating an abstraction as if it were a real thing.

Countermeasures: One way to combat reification is to constantly question our own assumptions and the language we use. Ask yourself, are you discussing an abstract concept as if it were a concrete entity? Also, try not to take things too literally; life's not a box of chocolates, even if Forrest Gump said so.

Key Takeaways: Remember to recognize when abstract concepts are being discussed as if they were concrete, and try not to fall into the trap of reification. And don't forget, there are no tickets to "Cloud Nine".

References:

Searle, J. R. (1995). The Construction of Social Reality. Free Press.

Epstein, S. (1994). Integration of the cognitive and the psychodynamic unconscious. American Psychologist, 49(8), 709–724. https://doi.org/10.1037/0003-066x.49.8.709

Rejection of Instruction

Definition: The Rejection of Instruction is a cognitive bias where individuals dismiss, ignore, or defy instructions or advice based on various factors like distrust, perceived superiority, or simple rebelliousness. It's the intellectual equivalent of "You're not my real dad!"

Description: This bias tends to crop up when people feel their autonomy or competence is threatened. It's a little bit like a teenage rebellion but in your brain. Just as some teenagers reject parental advice simply because it comes from their parents, some people reject instruction simply because it's instruction.

Examples: In politics, constituents might reject policies or directives from a party they oppose, even when these policies might benefit them. In religion, people may defy religious teachings or guidance based on their personal beliefs or prejudices. On social issues, advice from experts may be discarded simply due to skepticism or contrarian tendencies.

Impact: The Rejection of Instruction can lead to suboptimal decisions, inefficiency, and conflict. It's like stubbornly driving the wrong way even though your GPS is incessantly telling you to make a U-turn.

Related Biases: This bias is a close cousin to the Reactance Bias, where people do the opposite of what they're told because they perceive the instruction as a threat to their freedom.

Countermeasures: Keep an open mind and assess instructions based on their merits rather than your feelings towards the source. If your GPS tells you to turn around, consider that it might know the roads better than you do.

Key Takeaways: Don't dismiss instructions or advice out of hand just because you don't like being told what to do. Remember, even a broken clock is right twice a day.

References:

Burgoon, M., Alvaro, E., Grandpre, J., & Voulodakis, M. (2002). The Persuasion Handbook: Developments in Theory and Practice. Sage Publications.

Brehm, J. W. (1966). A Theory of Psychological Reactance. Academic Press.

Reminiscence Bump

Definition: The Reminiscence Bump refers to the cognitive bias where people recall more personal events from adolescence and early adulthood than from other periods of their lives. It's basically your brain's way of saying, "Ah, those were the days."

Description: This bias is believed to occur because adolescence and early adulthood are periods of many firsts: first love, first job, first existential crisis, etc. These novel experiences tend to be encoded more deeply into memory, making them easier to recall later in life.

Examples: In the political realm, you might remember the first president you voted for with more clarity than subsequent ones. In religious contexts, rituals or milestones experienced in early adulthood might be more memorable than later religious experiences. And socially, you're more likely to recall that time you went streaking through the quad in your college years rather than the countless mundane Tuesdays that followed in your 30s and 40s.

Impact: The Reminiscence Bump can affect how we perceive our past, leading to a rose-tinted view of our youth and potentially impacting our decisions and attitudes. If you find yourself saying "back in my day" a lot, you might be experiencing the Reminiscence Bump.

Related Biases: It's a fun cousin to the Peak-End Rule, where people judge experiences largely based on how they were at their peak and at their end, rather than based on the total sum or average of every moment of the experience.

Countermeasures: While it's impossible to stop the Reminiscence Bump from happening, being aware of it can help you understand that your memories may not be an accurate reflection of your past. Remember, the good old days weren't always good, and tomorrow ain't as bad as it seems.

Key Takeaways: Memories from your adolescence and early adulthood are likely to be more prominent, but it's essential to understand that your recollections might be somewhat skewed. Nostalgia ain't what it used to be.

References:

Rubin, D.C., Rahhal, T.A., & Poon, L.W. (1998). Things Learned in Early Adulthood are Remembered Best. Memory & Cognition, 26(1), 3-19.

Conway, M.A., & Haque, S. (1999). Overshadowing the Reminiscence Bump: Memories of a Struggle for Independence. Journal of Adult Development, 6(1), 35-44.

Representativeness Heuristic

Definition: The Representativeness Heuristic is the cognitive bias where people judge the probability or frequency of a hypothesis by considering how much the hypothesis resembles available data or the population. Essentially, it's the mind's own "looks like a duck, quacks like a duck" mechanism, with the added caveat that sometimes, it's not a duck.

Description: This bias leans on the principle of similarity – we tend to believe that similar things belong in the same category. While this often simplifies complex decision-making, it also leads us to overlook vital statistical information. The bias often takes center stage when base rates – the overall frequency of a category in the population – are neglected.

Examples: In politics, the heuristic may lead to stereotyping. If a person shares a few beliefs with a particular political group, they're often assumed to align completely with that group. In religion, an individual who displays some characteristics associated with a particular faith might be presumed to follow that religion. And in social issues, people might be lumped together based on a few superficial similarities, which can perpetuate stereotypes and inhibit understanding.

Impact: The representativeness heuristic can lead to poor decision-making and judgment. It can foster stereotypes, lead to overgeneralization, and limit our understanding of others. It can also contribute to the gambler's fallacy, the belief that past events can influence independent future events (e.g., "I've flipped tails five times, so the next one has to be heads!").

Related Biases: The Representativeness Heuristic is a close cousin of the Availability Heuristic (making decisions based on the information that is most readily available in our minds) and the Base Rate Fallacy (ignoring or underweighting base rate probabilities).

Countermeasures: To counter this bias, it's helpful to consider base rates and other statistical information consciously. Don't let appearances fool you; not every creature that quacks is a duck. Additionally, promoting diversity and avoiding stereotyping in all forms of interaction can go a long way in preventing the misuse of the Representativeness Heuristic.

Key Takeaways: The Representativeness Heuristic allows us to make quick judgments, but it's important to remember that it can be misleading. Ensure you have all the data before making a decision and beware of the potential for stereotyping and overgeneralization.

References:

Tversky, A., & Kahneman, D. (1974). Judgment under Uncertainty: Heuristics and Biases. Science, 185(4157), 1124-1131.

Kahneman, D., & Tversky, A. (1996). On the reality of cognitive illusions. Psychological Review, 103(3), 582–591.

Restraint Bias

Definition: Restraint Bias is the cognitive distortion where individuals overestimate their ability to control impulsive behavior. It's the intellectual equivalent of "eyes bigger than the stomach," but with a lack of self-control on the menu instead of food.

Description: This bias is about self-perception and the (mis)understanding of one's willpower. People afflicted with Restraint Bias often believe they can resist temptations better than they actually can. This overconfidence can lead to exposure to more temptation, thereby increasing the likelihood of succumbing to it.

Examples: In the realm of politics, a public servant might overestimate their ability to resist lobbying pressures. As for religion, a devotee may believe they can withstand religiously forbidden temptations, only to find themselves faltering. In the context of social issues, an individual might think they can resist peer pressure to engage in harmful behavior, yet they might end up doing it anyway.

Impact: Restraint Bias can lead to poor decision-making and an increased likelihood of indulging in undesirable or harmful behaviors. It also hampers self-regulation, making it harder to achieve personal goals, maintain health, or adhere to ethical standards.

Related Biases: Restraint Bias is a kissing cousin of the Overconfidence Effect, where individuals overestimate their abilities, and the Optimism Bias, where people believe they are less likely to experience negative events.

Countermeasures: Increasing self-awareness is key to mitigating Restraint Bias. Honest self-evaluation and regular feedback from others can help correct inflated perceptions of willpower. Also, it's good to remember that even Odysseus had himself tied to the mast to resist the Sirens' song. External safeguards can be lifesavers.

Key Takeaways: Restraint Bias is a tricky fellow. It makes you think you're stronger than you are, potentially leading you into temptation. To avoid this, honestly assess your willpower and don't be shy about using external safeguards when necessary.

References:

Nordgren, L. F., van Harreveld, F., & van der Pligt, J. (2009). The Restraint Bias: How the Illusion of Self-Restraint Promotes Impulsive Behavior. Psychological Science, 20(12), 1523–1528.

Quinn, P. D., & Fromme, K. (2010). Self-regulation as a protective factor against risky drinking and sexual behavior. Psychology of Addictive Behaviors, 24(3), 376–385.

Rhyme as Reason Effect

Definition: The Rhyme as Reason Effect is a cognitive bias where people perceive statements that rhyme as more truthful or accurate. Because "if it rhymes, it must be right," right?

Description: It's all about aesthetics here. When information is presented in a more pleasing manner (like a rhyme), it's often perceived as more accurate. This is primarily due to rhymed statements being easier to process and remember, thus leading people to believe they're more truthful.

Examples: The world of advertising is rife with this. "Red Bull gives you wings," anyone? In politics, catchy rhyming slogans sway voters ("Yes We Can"). As for religion, think of all those catchy hymns and verses. And let's not forget about social issues: "Click it or ticket" is a rhyming reminder to buckle up.

Impact: The Rhyme as Reason Effect can lead to uncritical acceptance of ideas if they're put into a catchy rhyme. It can skew perceptions and judgments, influence decision-making, and even be manipulated to spread misinformation.

Related Biases: It's a close relative of the Illusory Truth Effect (repeated statements are perceived as more truthful) and the Processing Fluency (easy-to-process statements are perceived as more accurate).

Countermeasures: Awareness of this bias is the first step. Always evaluate information critically, regardless of how catchy or aesthetically pleasing it is. And remember, just because it rhymes, doesn't make it right.

Key Takeaways: Rhyme as Reason Effect can make falsehoods seem true if they're catchy enough. Be aware of this bias and don't let a good rhyme sway your judgment.

References:

McGlone, M. S., & Tofighbakhsh, J. (2000). Birds of a feather flock conjointly (?): Rhyme as reason in aphorisms. Psychological Science, 11(5), 424-428.

Alter, A. L., & Oppenheimer, D. M. (2009). Uniting the tribes of fluency to form a metacognitive nation. Personality and Social Psychology Review, 13(3), 219-235.

Risk Compensation or Peltzman Effect

Definition: Risk Compensation or the Peltzman Effect is a cognitive bias that describes how people might take more risks if they feel safer due to safeguards or precautions. It's the "Oh, I have a seatbelt on; I can drive faster" kind of logic.

Description: It's a fascinating dance between safety and risk. When people feel protected, they sometimes take greater risks, assuming that these safeguards will prevent negative consequences. This behavior may potentially offset the benefits gained from the safeguards. The Peltzman Effect is named after Sam Peltzman, an economist who noted this pattern while studying automobile safety.

Examples: In politics, think about a politician feeling safe in their seat because of gerrymandering, leading to more controversial decisions. In religion, consider a person who commits sins believing that prayer will absolve them. On the social issues front, someone may ignore social distancing guidelines because they're wearing a mask.

Impact: This effect can lead to risky behaviors and decision-making that can negate the benefits of safety measures. It can also affect public policy and societal norms, as well as personal choices that could have serious repercussions.

Related Biases: The Peltzman Effect is linked with Optimism Bias (overestimating positive outcomes) and Illusion of Control (overestimating one's control over events).

Countermeasures: Self-awareness and education are critical. Understand the purpose of safety measures and the risks of not adhering to them. Policies should also consider potential compensatory behaviors when implementing safety measures.

Key Takeaways: Feeling safer can lead to riskier behavior — a phenomenon known as Risk Compensation or the Peltzman Effect. Recognizing this bias is the first step to ensuring that safety measures serve their intended purpose.

References:

Peltzman, S. (1975). The effects of automobile safety regulation. Journal of political economy, 83(4), 677-725.

Wilde, G. J. (1982). The theory of risk homeostasis: implications for safety and health. Risk analysis, 2(4), 209-225.

Rosy Retrospection

Definition: Rosy retrospection refers to the cognitive bias that leads us to recall past events more fondly and positively than they actually were. It's the mental equivalent of looking at the world through rose-tinted glasses, but only when glancing backward.

Description: Ah, the good old days, when everything was better, right? Not exactly. Rosy retrospection is an offshoot of memory bias, where our brain polishes past experiences, making them look shiny and more pleasant in hindsight. It's like having a mental Instagram filter that glosses over unpleasantness and amps up the highlights of the past.

Examples: In politics, you'll see rosy retrospection every time someone fondly remembers the "golden age" of a certain leadership, conveniently forgetting the problems of that era. In religion, it manifests as longing for a mythical golden age. In social issues, it's visible when older generations grumble about how manners, music, or morals were "better in their day."

Impact: This bias can distort our perceptions and decision-making. It can make us resistant to change, cling to outdated traditions, or repeat mistakes because we misremember the outcomes. By fostering nostalgia, it can impede progress and hinder growth.

Related Biases: Confirmation bias (focusing on information that confirms our pre-existing beliefs), Hindsight bias (misremembering our predictions to fit actual outcomes), and Nostalgia bias (believing the past was better than it was).

Countermeasures: Critical thinking is crucial. Try to remember events as objectively as possible. Fact-checking your own memories, using diaries or photos, can help. When making decisions, rely on documented facts and evidence, not just memory.

Key Takeaways: Rosy retrospection makes us remember the past as better than it was. It can distort our decision-making and make us resistant to change. Counter it by thinking critically and fact-checking your memories.

References:

Mitchell, T. R., Thompson, L., Peterson, E., & Cronk, R. (1997). Temporal adjustments in the evaluation of events: The "rosy view". Journal of Experimental Social Psychology, 33(4), 421–448.

Fredrickson, B. L. (2000). Extracting meaning from past affective experiences: The importance of peaks, ends, and specific emotions. Cognition & Emotion, 14(4), 577–606.

Satisficing

Definition: Satisficing is a decision-making strategy that aims for a satisfactory or "good enough" outcome, rather than the optimal solution. It's the art of accepting "meh" instead of hunting relentlessly for perfection.

Description: The term "satisfice" combines "satisfy" and "suffice," and was coined by economist and Nobel laureate Herbert Simon. It's the behavioral economics equivalent of settling for a decent restaurant because you're too hungry to keep looking for the perfect one. This bias arises due to limited resources like time, information, or mental capacity, which prohibits thorough analysis of all options.

Examples: In politics, voters might select a candidate they find "good enough" rather than thoroughly researching all candidates. In religion, someone might choose a belief system that appears satisfactory without exploring all religious philosophies in-depth. In social issues, someone might settle on a viewpoint that seems reasonable rather than critically examining all perspectives.

Impact: Satisficing can result in sub-optimal decisions. However, it also reduces the time and effort spent on decision-making. While it can be efficient in some situations, it may also lead to missed opportunities and less-than-ideal outcomes in others.

Related Biases: The Paradox of Choice (where having too many options leads to decision paralysis) and Decision Fatigue (where decision quality deteriorates after making many decisions).

Countermeasures: Awareness is key. Know when you're satisficing and make conscious decisions about when it's appropriate. In critical decisions, strive for optimal outcomes. Employ decision-making tools and frameworks to facilitate comprehensive analysis.

Key Takeaways: Satisficing is settling for "good enough" rather than the best. While efficient, it can lead to less-than-optimal decisions. Distinguish between decisions that require optimal outcomes and those where satisficing is suitable.

References:

Simon, H. A. (1956). Rational choice and the structure of the environment. Psychological review, 63(2), 129–138.

Schwartz, B. (2004). The paradox of choice: Why more is less. New York: Ecco.

Scarcity Principle

Definition: The scarcity principle refers to the cognitive bias that makes us perceive something as more valuable when it is less available or seems to be in high demand. It's the classic "limited-time offer" ploy – psychological manipulation in its most tempting form.

Description: Ever get that pang of FOMO (fear of missing out) when something is "only available for a limited time" or "while supplies last"? Well, meet the culprit: the scarcity principle. It exploits our innate fear of losing out on opportunities and our perception that scarce things are often more valuable. It's like your brain's version of "you don't know what you've got till it's gone."

Examples: Politics uses the scarcity principle when framing issues as urgent problems that need immediate attention. In religion, scarcity can be used to reinforce the special status of certain artifacts or practices. In social issues, it can create false dilemmas where two options are presented as the only choices, thereby increasing their perceived value.

Impact: This bias can lead us to make hasty, impulsive decisions, often guided more by emotion than rational thought. It can induce a sense of urgency that overrides our better judgement, leading us to overlook red flags, ignore better alternatives, or pay more than necessary.

Related Biases: Loss aversion (preferring to avoid losses than to gain equivalent things), FOMO (fear of missing out), and The bandwagon effect (believing or doing things because many other people do the same).

Countermeasures: Slow down. Take time to evaluate options and consider alternatives. Ask yourself: "Would I want this if there were plenty of it?" Remember that scarcity is often artificially created for marketing purposes.

Key Takeaways: The scarcity principle makes us see scarce things as more valuable, leading to hasty and impulsive decisions. Counter it by slowing down, evaluating options, and not being swayed by artificially created scarcity.

References:

Cialdini, R. B. (2001). Influence: Science and practice (4th ed.). Allyn & Bacon.

Lynn, M. (1991). Scarcity effects on value: A quantitative review of the commodity theory literature. Psychology & Marketing, 8(1), 43–57.

Scope Insensitivity

Definition: Scope insensitivity is a cognitive bias where the valuation of a problem or a life is not proportionate to its scope. If saving one life seems worth $10,000, the notion of saving 100 lives doesn't necessarily translate to a $1,000,000 willingness to pay in our minds, even though, mathematically, it should.

Description: This is the bias that makes you think, "Why would I donate $100 to save 100 rainforest acres when I can donate $10 to save one?" It's as if our brain hits a saturation point where it can't grasp large quantities – a kind of numerical numbing, if you will. In essence, it's our brain's version of "well, that escalated quickly."

Examples: In politics, scope insensitivity can lead to the underestimation of the scale of problems, leading to insufficient resource allocation. In religious context, the magnitude of moral or ethical concerns may be inappropriately assessed due to this bias. Social issues, especially those involving large-scale suffering or injustice, often fall victim to scope insensitivity, with people feeling the same level of concern for small and large groups.

Impact: Scope insensitivity can lead to flawed decision making, inappropriate allocation of resources, and undervaluation of large-scale issues. It can lead to a lack of proportionality in our judgments and decisions, where the scale of our response does not match the scale of the problem.

Related Biases: The identifiable victim effect, where people show greater sympathy for identifiable victims than statistical victims; and compassion fade, where empathy decreases as the number of people in need increases.

Countermeasures: Consciously work to maintain proportionality in your decisions and judgments. When considering issues or problems, take the time to understand their scale and adjust your response accordingly. Try to break down large numbers into smaller, more understandable figures.

Key Takeaways: Scope insensitivity is the bias that makes us unable to keep our valuation of problems or lives proportional to their scale. To counteract it, we need to consciously maintain proportionality in our judgments and decisions.

References:

Desvousges, W. H., Johnson, F. R., Dunford, R. W., Boyle, K. J., Hudson, S. P., & Wilson, K. N. (1993). Measuring natural resource damages with contingent valuation: Tests of validity and reliability. In Contingent Valuation: A Critical Assessment (pp. 91-164). Elsevier.

Slovic, P. (2007). “If I look at the mass I will never act”: Psychic numbing and genocide. Judgment and Decision Making, 2(2), 79-95.

Selective Exposure

Definition: Selective exposure is a cognitive bias where individuals favor information that reinforces their pre-existing views while avoiding contradictory information. It's like choosing to eat only chocolate and ignoring broccoli, despite knowing that a balanced diet would be healthier.

Description: Selective exposure is when we slyly avoid any information that could mess up our nicely organized mental furniture. We cherry-pick our way through life, opting for the familiar, the affirming, and the ego-stroking. Let's be honest, it's much more comforting to stick to our well-trodden thought patterns than to traipse through the wild jungles of alternative perspectives.

Examples: A staunch political partisan might only watch news channels that align with their own political beliefs, deliberately ignoring outlets with opposing viewpoints. Or in the realm of religion, someone may strictly adhere to their own faith's teachings, avoiding scholarly works or discussions that challenge their beliefs. When dealing with social issues, people might choose to interact only with those who hold similar viewpoints, avoiding diverse opinions that could disrupt their worldview.

Impact: This bias can limit critical thinking, create echo chambers, and increase polarization in society. It can make us blind to alternative perspectives and make our own beliefs rigid and unyielding. This lack of intellectual flexibility can prevent societal growth and foster conflict.

Related Biases: Confirmation bias, where we search for and interpret information that confirms our preconceptions;

and cognitive dissonance, where we experience discomfort when we hold two contradictory beliefs or values at the same time.

Countermeasures: Actively seek out different perspectives and consider the merits of views that differ from your own. Engage in discussions with people who have diverse opinions. Be aware that no one, including you, has a monopoly on truth or wisdom.

Key Takeaways: Selective exposure is the tendency to favor information that aligns with our pre-existing beliefs and avoid information that challenges them. To counter it, we need to consciously seek out diverse perspectives and remain open to alternative viewpoints.

References:

Frey, D. (1986). Recent research on selective exposure to information. Advances in experimental social psychology, 19, 41-80.

Hart, W., Albarracín, D., Eagly, A. H., Brechan, I., Lindberg, M. J., & Merrill, L. (2009). Feeling validated versus being correct: A meta-analysis of selective exposure to information. Psychological bulletin, 135(4), 555.

Self-Consistency Bias

Definition: Self-consistency bias is the mental hocus-pocus that leads us to believe our attitudes and behaviors are more consistent than they actually are. It's like our mind is playing a constant game of "Connect the Dots," where it conveniently ignores the dots that don't align with our self-perception.

Description: In the grand theatre of our minds, we fancy ourselves as the stalwart lead who never falters, consistently following the script we've written for ourselves. But here's the reality check: We are more akin to improvisational actors, constantly adjusting to changing circumstances and then conveniently forgetting about these adjustments. Self-consistency bias ensures we maintain a sense of coherence and continuity, even when faced with the embarrassing truth of our inconsistency.

Examples: If a deeply religious individual occasionally doubts their faith but downplays or forgets these moments of skepticism, they're experiencing self-consistency bias. Politically, someone may have shifted their stance over time but believes they've always held their current position. In social issues, people might deny ever having held prejudiced views, despite evidence to the contrary.

Impact: Self-consistency bias can prevent us from learning and growing. We might fail to recognize the evolution of our perspectives, or worse, stick to detrimental beliefs or behaviors in the name of consistency. It can also lead to misunderstanding and conflict when we project our own perceived consistency onto others, expecting them to adhere to our constructed image of their character.

Related Biases: Confirmation bias, where we favor information that confirms our existing beliefs; and self-serving bias, where we perceive ourselves in an overly favorable manner.

Countermeasures: Regularly reflect on your beliefs and attitudes to recognize and acknowledge changes. Accept that change and growth are a part of life, and it's okay not to be consistent all the time. Practice empathy and understanding when dealing with others' inconsistencies.

Key Takeaways: Self-consistency bias is our mental trick of maintaining a coherent self-image by overestimating the consistency of our attitudes and behaviors. Embrace change and growth, and don't be afraid to update your mental script.

References:

Gawronski, B., & Strack, F. (Eds.). (2012). Cognitive consistency: A fundamental principle in social cognition. Guilford Press.

Festinger, L. (1957). A theory of cognitive dissonance. Stanford University Press.

Self-Enhancement Bias

Definition: The self-enhancement bias is our mind's sneaky little habit of viewing ourselves more favorably than is objectively warranted. It's like wearing rose-tinted glasses, but only when looking in the mirror.

Description: Our egos, bless their heart, love a good pat on the back. To secure such ego-stroking, they tend to lean on the self-enhancement bias. This involves overestimating our good qualities and achievements while downplaying or outright ignoring our not-so-glorious moments. The result? An inflated sense of self that may feel like a cozy cocoon but, alas, might not be a true reflection of reality.

Examples: An example of this bias in the religious domain could be believing that one is more virtuous or pious than others. In politics, a candidate may overestimate their popularity and chances of winning an election. On social issues, someone might believe they're more open-minded and less prejudiced than most people, even when their actions indicate otherwise.

Impact: While a touch of self-enhancement can boost our self-esteem and resilience, too much can make us delusional, hampering personal growth. It can also harm relationships, as people tend to dislike those who overestimate their positive traits, labelling them as arrogant or narcissistic.

Related Biases: Overconfidence bias, where we place too much faith in our knowledge or abilities; and self-serving bias, where we take credit for success but blame others or external factors for failures.

Countermeasures: Cultivate self-awareness and humility. Seek feedback from others and be open to constructive criticism. Don't fall for the allure of self-flattery but strive for an accurate self-perception.

Key Takeaways: The self-enhancement bias can serve as an ego-boosting comfort blanket, inflating our self-image to a level that may border on delusion. To counter this, embrace humility, seek genuine feedback, and commit to seeing yourself as you truly are, warts and all.

References:

Alicke, M. D., & Sedikides, C. (Eds.). (2011). Handbook of self-enhancement and self-protection. Guilford Press.

Sedikides, C., & Gregg, A. P. (2008). Self-enhancement: Food for thought. Perspectives on Psychological Science, 3(2), 102-116.

Self-Handicapping Bias

Definition: The self-handicapping bias is our knack for creating obstacles and excuses to use as a handy scapegoat in case we fail. It's like sabotaging your own race, then blaming the rocky path you laid out for yourself.

Description: This bias is a preemptive strike against potential failure. By creating or claiming impediments, we protect our self-esteem. After all, it's easier to point fingers at an 'unfortunate situation' than to accept we might not have cut the mustard. This way, we get a get-out-of-jail-free card for our shortcomings, while successes seem even grander against the self-imposed odds.

Examples: For instance, in religion, a person may claim they can't meditate due to an 'overactive mind', thereby excusing any lack of spiritual growth. In politics, a candidate may under-prepare for a debate, blaming a 'lack of time', to have an excuse if they perform poorly. And in social issues, people may claim they can't help being prejudiced because 'it's how they were raised', shifting blame from their own actions to their upbringing.

Impact: While it can shield self-esteem in the short term, self-handicapping can be destructive in the long run. It prevents personal growth and learning, fosters procrastination, and can harm reputation and relationships, as others may see through the excuses and view the person as lacking accountability.

Related Biases: Self-serving bias, where people attribute successes to themselves and failures to external factors, and confirmation bias, where people seek information that confirms their existing beliefs or excuses.

Countermeasures: The first step is awareness and recognition of this behavior. Then, fostering a growth mindset can help, where effort and learning are valued more than outcomes. Embracing failures as opportunities for learning and growth rather than threats to self-esteem can also be beneficial.

Key Takeaways: The self-handicapping bias is our inclination to sabotage our own success to safeguard our self-image. While it may provide temporary comfort, it can impede personal growth and accountability. Countering it involves fostering a growth mindset and embracing failures as stepping stones to success.

References:

Berglas, S., & Jones, E. E. (1978). Drug choice as a self-handicapping strategy in response to noncontingent success. Journal of Personality and Social Psychology, 36(4), 405.

Zuckerman, M., & Tsai, F. F. (2005). Costs of self-handicapping. Journal of personality, 73(2), 411-442.

Self-Perception Bias

Definition: Self-perception bias is the act of being the protagonist in your own story while being unable to read your lines. It's our tendency to inaccurately assess our own behaviors and characteristics, often for the better, kind of like squinting in the mirror after forgetting your glasses.

Description: This bias often stems from a lack of self-awareness and a desire to maintain a positive self-image. You know how when you see a video of yourself and think, "Do I really sound like that?" That's self-perception bias in action. It's like being the lead singer in your own life's band, but forgetting the words to your song.

Examples: In religion, one might overestimate their commitment to their faith's tenets. In politics, an official might perceive their policy choices as more popular than they actually are. And in social contexts, someone may see themselves as a brilliant conversationalist, while others might find them a tad too chatty.

Impact: The self-perception bias can lead to overconfidence and an inflated sense of competence or morality. This could result in poor decisions, relationship conflicts, or even public embarrassment if our self-perception greatly deviates from how others see us. It's a bit like stepping on stage thinking you're a rockstar, only to find you can't strum a guitar.

Related Biases: Overconfidence effect, where one's confidence in their judgments is greater than the objective accuracy of those judgments, and self-serving bias, where successes are attributed to oneself and failures to external factors.

Countermeasures: Increasing self-awareness can help mitigate this bias. Seeking out feedback from trusted sources can also provide a reality check. Humility is the key; it's recognizing you're not the rockstar you think you are - unless you can actually play a decent riff on that guitar.

Key Takeaways: Self-perception bias can often lead to an inaccurate assessment of our own abilities and traits. It can result in overconfidence and poor decision making. To counteract this bias, increase self-awareness, seek out feedback, and embrace humility.

References:

Bem, D. J. (1967). Self-perception: An alternative interpretation of cognitive dissonance phenomena. Psychological review, 74(3), 183.

Heider, F. (1958). The Psychology of Interpersonal Relations. John Wiley & Sons.

Self-Relevance Effect

Definition: Self-relevance effect is the cognitive bias equivalent of a child who thinks the universe revolves around them. It's our tendency to perceive and interpret information as more important if it's personally relevant to us.

Description: This bias operates on the principle of "me, myself, and I." It's driven by our survival instincts – we're biologically wired to prioritize information that could directly affect our well-being. It's like having a mental spotlight that constantly shines on anything that screams, "This matters to me!"

Examples: In religion, believers may emphasize interpretations of scripture that align with their personal values or life experiences. In politics, a person may pay more attention to policies that directly impact them, such as tax laws for their income bracket. In social issues, someone might prioritize news stories about their hometown or profession.

Impact: While self-relevance can be useful for sifting through heaps of information, it can also lead to a skewed perspective, neglecting broader or more abstract issues. Picture being so focused on the ingredients of your morning coffee that you ignore the kitchen fire.

Related Biases: Confirmation bias, where we favor information that confirms our existing beliefs, and the not invented here bias, where we dismiss ideas simply because they come from outside our personal or cultural group.

Countermeasures: Encourage a more balanced view by actively seeking out diverse information and perspectives.

Empathy and understanding other viewpoints can also help. And remember, despite what your instincts tell you, not everything is about you.

Key Takeaways: Self-relevance effect leads us to prioritize information that we see as personally relevant, potentially ignoring important broader issues. It can be counteracted by seeking diverse information and cultivating empathy.

References:

Rogers, T. B., Kuiper, N. A., & Kirker, W. S. (1977). Self-reference and the encoding of personal information. Journal of Personality and Social Psychology, 35(9), 677.

Symons, C. S., & Johnson, B. T. (1997). The self-reference effect in memory: a meta-analysis. Psychological bulletin, 121(3), 371.

Self-Reference Effect

Definition: The self-reference effect is a cognitive bias that highlights our tendency to better remember information when we can relate it to ourselves. It's like our brain's version of a "stick-it-to-the-fridge" note, but the fridge is us.

Description: This bias operates like a highly effective sticky note in our mind. When information is connected to our personal experiences or aspects of our identity, we're more likely to recall it. After all, nothing spices up the bland casserole of facts like a pinch of self-interest.

Examples: In religion, one might better recall parables that align with personal experiences. In politics, if a policy directly affects your job, you're more likely to remember the politician who proposed it. For social issues, it's easier to remember statistics about pollution if you, yourself, have been affected by poor air quality.

Impact: While the self-reference effect can enhance learning and memory, it can also lead to a narrow worldview. Imagine remembering only the conversations where you did the talking. Your understanding would be pretty one-sided, right?

Related Biases: This bias is a cousin of the self-relevance effect and an occasional partner-in-crime with confirmation bias.

Countermeasures: Diversifying sources of information, focusing on empathy, and actively seeking perspectives outside of one's own experiences can help mitigate this bias. Don't forget: there's a whole universe beyond your navel.

Key Takeaways: The self-reference effect enhances our memory of information that we can connect to ourselves, but it can also narrow our perspectives. Counteract this by seeking diverse experiences and viewpoints.

References:

Rogers, T. B., Kuiper, N. A., & Kirker, W. S. (1977). Self-reference and the encoding of personal information. Journal of Personality and Social Psychology, 35(9), 677.

Symons, C. S., & Johnson, B. T. (1997). The self-reference effect in memory: a meta-analysis. Psychological bulletin, 121(3), 371.

Semmelweis Reflex

Definition: The Semmelweis Reflex is the knee-jerk rejection of new information or ideas, especially if they conflict with established norms, beliefs, or paradigms. It's the cognitive equivalent of crossing one's arms and declaring, "We've always done it this way!"

Description: The bias is named after Ignaz Semmelweis, a Hungarian physician who recommended that doctors wash their hands to prevent spreading disease. Though it sounds pretty elementary to us today, in the 19th century, this idea was about as popular as a skunk at a garden party. This bias highlights our tendency to hold on to old ways like a dog with a favorite chew toy, even when presented with evidence that change might be beneficial.

Examples: Within religion, this reflex can be seen when individuals dismiss progressive interpretations of traditional teachings. In politics, it can be observed in the rejection of transformative policies in favor of "the way things have always been done". Social issues are rife with examples as well, such as opposition to changing social norms around gender or race.

Impact: This bias stifles innovation and can impede the adoption of beneficial practices. The Semmelweis Reflex might have you clinging to the comfortable past while the world leaps forward without you.

Related Biases: Confirmation bias and status quo bias are close relatives of the Semmelweis Reflex.

Countermeasures: Cultivate an open mind. Welcome new ideas for consideration, even if they seem strange or challenging. Encourage dialogue and constructive debate. Remember, at one point, sliced bread was a revolutionary idea.

Key Takeaways: The Semmelweis Reflex describes the automatic dismissal of new or conflicting information. To counteract it, embrace new ideas and perspectives—after all, change is the only constant.

References:

Ernst, E. (2014). Semmelweis's early studies on hand hygiene: a tale of heroism and tragedy. Journal of hospital infection, 86(3), 174-178.

Duska, R. (1975). The Semmelweis reflex. BioScience, 25(9), 561-563.

Shared Information Bias

Definition: Shared Information Bias is the tendency to discuss, consider, and overemphasize information that everyone knows while neglecting critical nuggets of wisdom only a few hold. It's like ignoring the ingredients list on a cake mix box and ending up with a brick instead of a fluffy dessert because everyone knows how to make a cake, right?

Description: This bias often appears in group decisions and discussions, where the known and familiar is more comfortable to talk about than the unique insights of individuals. It's akin to everyone discussing the color of the drapes while the house is on fire.

Examples: In religious groups, commonly shared beliefs might overshadow unique interpretations or questioning. Similarly, in politics, a group of like-minded individuals may keep reaffirming each other's views, neglecting to consider a unique perspective from a newcomer. As for social issues, the voices of marginalized groups might be drowned out in conversations dominated by the majority.

Impact: The Shared Information Bias can lead to poor decision-making, a lack of innovation, and groupthink. It's the equivalent of everyone agreeing the ship is unsinkable while dismissing that lone voice mentioning something about an iceberg.

Related Biases: This bias is related to the Confirmation Bias, where we focus on information that confirms our existing beliefs, and Groupthink, where conformity in a group can lead to irrational decisions.

Countermeasures: Encourage the sharing of unique information in group settings, foster an environment where dissent is welcomed, and remind yourself that common knowledge isn't always the most useful knowledge.

Key Takeaways: The Shared Information Bias illustrates how groups can focus too much on common knowledge at the expense of unique insights. To mitigate this, emphasize the sharing of diverse perspectives and encourage critical thinking over conformity.

References:

Stasser, G., & Titus, W. (1985). Pooling of unshared information in group decision making: Biased information sampling during discussion. Journal of personality and social psychology, 48(6), 1467.

Wittenbaum, G. M., Hubbell, A. P., & Zuckerman, C. (1999). Mutual enhancement: Toward an understanding of the collective preference for shared information. Journal of personality and social psychology, 77(5), 967.

Simulation Heuristic

Definition: The Simulation Heuristic is our mental tendency to judge the likelihood of an event based on how easily we can imagine or simulate it in our minds. It's like when you conclude that you're more likely to win the lottery because you've spent the entire morning imagining your victory press conference.

Description: This heuristic, or mental shortcut, often takes center stage when we deal with uncertainty. It's the reason why we feel more scared of plane crashes than car accidents, despite the former being statistically less frequent. Our brains are just big fans of compelling narratives, even if they are not the most statistically probable.

Examples: Within religious contexts, tales of miraculous healings or divine interventions could be accepted as more plausible simply because they make for a good mental movie. In politics, a candidate may seem more likely to win because we can easily imagine their victory speech. For social issues, a sensational but unlikely event could gain more public attention and concern than more prevalent, but less dramatic issues.

Impact: The Simulation Heuristic can lead to misjudgments about the likelihood of events, causing us to focus on the dramatic or easily-imagined at the expense of the statistically probable. This bias can skew our understanding of reality, like being convinced that your odds of being attacked by a shark are astronomically higher after watching "Jaws."

Related Biases: This heuristic is related to the Availability Heuristic, where we base judgments on the information that readily comes to mind, and the Representativeness Heuristic, where we judge probabilities based on perceived similarities.

Countermeasures: Try to base decisions on objective information and statistical data whenever possible. Remind yourself that the ease with which you can imagine an event doesn't necessarily correlate with its likelihood.

Key Takeaways: The Simulation Heuristic reflects our tendency to mistake mental drama for probability. To mitigate its impact, strive to separate vivid imagination from objective reality, particularly when dealing with uncertainties.

References:

Kahneman, D., & Tversky, A. (1982). The simulation heuristic. In Judgment under uncertainty: Heuristics and biases (pp. 201-208). Cambridge University Press.

Tversky, A., & Kahneman, D. (1973). Availability: A heuristic for judging frequency and probability. Cognitive psychology, 5(2), 207-232.

Sleeper Effect

Definition: The Sleeper Effect is a psychological phenomenon where information from unreliable sources initially has low persuasive impact but increases over time as we forget the source's unreliability. Think of it as the mental equivalent of aging wine, but in this case, the vintage is dubious, and the vintner has a reputation for stretching the truth.

Description: This bias arises from our minds' knack for playing tricks on us. When we first receive information from a source we deem unreliable, we're likely to discount it. But as time goes by, the association between the information and its source weakens in our memories, allowing the message to stand on its own. Like a dubious wine you've had in your cellar for a while, you forget that it was never that great to start with and decide to give it a go.

Examples: In religious contexts, a controversial prophecy might gain acceptance over time as people forget the unreliable origins of the message. In politics, a rumor about a candidate may initially be dismissed due to a disreputable source, but gain traction as people forget the source and only remember the rumor. Socially, gossip and hearsay can become 'common knowledge' over time, regardless of the reliability of the original source.

Impact: The Sleeper Effect can contribute to the spread of misinformation and false beliefs, influencing our perceptions and decisions. It's like accidentally drinking spoiled wine because you've forgotten its questionable origins, and no one likes the hangover from that.

Related Biases: Source Amnesia is closely related, referring to our tendency to remember information without recalling where or how we learned it.

Countermeasures: To mitigate the Sleeper Effect, make a conscious effort to evaluate the credibility of sources and maintain that association over time. Or in other words, remember where you bought that wine and why you didn't drink it immediately.

Key Takeaways: The Sleeper Effect represents the dangerous potential of information from unreliable sources gaining credibility over time. Remember, not all that ages is fine wine. Sometimes, it's just old grape juice.

References:

Kumkale, G. T., & Albarracín, D. (2004). The sleeper effect in persuasion: A meta-analytic review. Psychological Bulletin, 130(1), 143–172.

Pratkanis, A. R., Greenwald, A. G., Leippe, M. R., & Baumgardner, M. H. (1988). In search of reliable persuasion effects: III. The sleeper effect is dead. Long live the sleeper effect. Journal of Personality and Social Psychology, 54(2), 203–218.

Social Attribution Bias

Definition: Social Attribution Bias is the tendency to perceive others' actions as indicative of their character or intentions, rather than attributing their behavior to external, situational factors. It's essentially the social version of assuming the worst-case scenario without considering the circumstances.

Description: We all play Sherlock Holmes in our day-to-day interactions, trying to deduce what's going on in others' minds. But unlike the fictional detective, we often miss crucial contextual clues and make hasty assumptions about people's character. This bias suggests we often overestimate dispositional influences (like personality) while underestimating situational ones (like environmental pressures).

Examples: In religion, someone's lack of faith might be attributed to their perceived moral failings rather than their personal experiences or lack of evidence. In politics, a leader's unpopular decision could be seen as a sign of incompetence or malice, overlooking the possibility of difficult circumstances or pressures behind the scenes. On a social level, if a friend doesn't return your call, it's easy to assume they're being thoughtless, instead of considering they might be tied up with problems you're unaware of.

Impact: Social Attribution Bias can foster misunderstandings and conflict in interpersonal relationships and broader societal issues. It's like assigning blame for a burnt dinner without considering the faulty oven or the unexpected phone call that distracted the cook.

Related Biases: Fundamental Attribution Error and Actor-Observer Bias are closely related phenomena, reflecting our inclination to favor dispositional explanations for others' behavior and situational ones for our own.

Countermeasures: Encouraging empathy, perspective-taking, and mindfulness can help counteract this bias. Before making a judgment, try to understand the situation from the other person's perspective—walk a mile in their shoes, or at least glance at the path they're treading on.

Key Takeaways: The Social Attribution Bias serves as a reminder that our judgments about others often overlook the influence of situational factors. Remember, not everything is as it seems, and not every burnt dinner is a sign of a terrible cook.

References:

Jones, E. E., & Nisbett, R. E. (1971). The actor and the observer: Divergent perceptions of the causes of behavior. In E. E. Jones et al. (Eds.), Attribution: Perceiving the causes of behavior. General Learning Press.

Malle, B. F. (2006). The actor-observer asymmetry in attribution: A (surprising) meta-analysis. Psychological Bulletin, 132(6), 895–919.

Social Comparison Bias

Definition: Social Comparison Bias is the tendency to compare ourselves to others, generally in ways that put us in a more favorable light. It's like holding a beauty contest where we've bribed the judges and we're the only contestants.

Description: Humans have an uncanny knack for turning life into an endless series of competitions, even when no official contest is underway. The Social Comparison Bias reflects our compulsion to measure our worth against others', often skewing the comparison in our favor. This bias can be divided into two types: upward comparison, when we compare ourselves with someone we perceive as superior, and downward comparison, when we compare ourselves to those we see as inferior.

Examples: In religion, a practitioner may perceive their faith as more authentic or fulfilling than others, boosting their spiritual self-esteem. In politics, supporters of a particular party might downplay their party's shortcomings by constantly pointing out the flaws of rival parties. On social media, it's easy to feel superior when our latest post garners more likes than a friend's post.

Impact: While this bias can boost self-esteem, it can also fuel dissatisfaction, resentment, and division. In the quest for validation, we may lose sight of our individual strengths and overlook the value of diversity.

Related Biases: Superiority Bias, where we view ourselves as better than average, is a direct offshoot of this bias. The Halo Effect and Horns Effect can also come into play, as our overall impression of someone can influence how we view their individual attributes.

Countermeasures: Cultivate self-awareness and self-acceptance. Acknowledge your unique strengths and limitations, independent of others. Practice gratitude for what you have, rather than focusing on what others have that you don't.

Key Takeaways: The Social Comparison Bias illustrates our tendency to measure our self-worth against others', often with a bit of creative accounting. Remember that each of us is a unique individual with our own set of strengths and weaknesses, and comparison often leads to discontent.

References:

Festinger, L. (1954). A theory of social comparison processes. Human relations, 7(2), 117-140.

Suls, J., Martin, R., & Wheeler, L. (2002). Social comparison: Why, with whom, and with what effect?. Current directions in psychological science, 11(5), 159-163.

Social Desirability Bias

Definition: Social Desirability Bias is the tendency for people to present themselves in a manner that will be viewed favorably by others. It's the psychological equivalent of Photoshop, only there's no "undo" button.

Description: This bias is rooted in our desire to fit in, avoid social disapproval, and maintain a positive self-concept. It often leads to over-reporting of good behavior and under-reporting of undesirable behavior. Think of it as society's "filter" feature, making everyone appear just a bit better than reality.

Examples: Consider a religious leader who advocates for purity but secretly indulges in vices. Or a politician who exaggerates their accomplishments and glosses over their failures. In social settings, this bias leads to a culture where everyone's life seems perfect on Instagram, even though we all know the reality is different.

Impact: This bias can distort self-reports in research, muddle public discourse, and perpetuate unrealistic standards of behavior. It can also create a dissonance between one's public image and private reality, leading to internal conflict and stress.

Related Biases: The Hawthorne Effect is related, as it involves people altering their behavior due to awareness of being observed. Impression Management, too, where individuals try to control the impressions others form of them.

Countermeasures: Develop self-awareness and integrity, valuing authenticity over social approval. In research, use anonymous data collection methods to mitigate this bias.

Key Takeaways: Social Desirability Bias reminds us that all that glitters is not gold. It's essential to question, critically and compassionately, the gap between appearances and reality, in ourselves and others.

References:

Fisher, R. J. (1993). Social Desirability Bias and the Validity of Indirect Questioning. Journal of Consumer Research, 20(2), 303-315.

van de Mortel, T. F. (2008). Faking it: social desirability response bias in self-report research. Australian Journal of Advanced Nursing, The, 25(4), 40.

Social Loafing

Definition: Social Loafing is the phenomenon where individuals put less effort into a task when they're part of a group. It's the psychological equivalent of "riding on coattails," only in this case, the coattails often lead to the land of mediocrity.

Description: This bias emerges because people believe their individual contribution is less noticeable in a group setting. The responsibility gets diffused, and everyone assumes someone else will pick up the slack. It's like a group project in school where you hoped the smart kid would carry the load.

Examples: In the realm of politics, think about a committee tasked with drafting new regulations. Everyone might lean on the most diligent member, leading to less diverse input and poorer outcomes. In religious congregations, members may rely too heavily on clergy for outreach, neglecting their own responsibilities.

Impact: Social Loafing can undermine group productivity, reduce overall output, and foster resentment among team members. It may also stifle individual initiative and creativity.

Related Biases: The Bystander Effect, where individuals are less likely to offer help to a victim when other people are present, and Free Rider Problem, where individuals take advantage of a common resource without contributing to it.

Countermeasures: Encourage individual accountability, set clear expectations, and recognize individual contributions. Breaking tasks into identifiable components can also help.

Key Takeaways: Social Loafing is a reminder that while "teamwork makes the dream work," we must also foster individual effort and responsibility within the team.

References:

Karau, S. J., & Williams, K. D. (1993). Social loafing: A meta-analytic review and theoretical integration. Journal of Personality and Social Psychology, 65(4), 681-706.

Latané, B., Williams, K., & Harkins, S. (1979). Many hands make light the work: The causes and consequences of social loafing. Journal of Personality and Social Psychology, 37(6), 822-832.

Stereotype Boost and Stereotype Lift

Definition: Stereotype Boost and Stereotype Lift are phenomena where individuals perform better when aware of positive stereotypes about their social group. In essence, it's when someone says, "I come from a long line of geniuses," and subsequently solves a Rubik's cube in record time.

Description: Stereotype Boost refers to an individual's enhanced performance after exposure to a positive stereotype about their in-group. Stereotype Lift refers to improved performance due to awareness of a negative stereotype about an out-group. It's like running faster after hearing that your group is known for speed, or running faster still knowing the group you're not part of isn't.

Examples: In politics, a person may rise to leadership roles fueled by positive stereotypes about their gender or race. In religious contexts, a clergy member may excel in pastoral duties, buoyed by the belief that people of their faith are empathetic and wise.

Impact: While these biases can enhance performance, they also uphold stereotyping, which can perpetuate harmful narratives and inequalities. They reinforce the idea that group-based expectations should shape individual actions, creating pressure to conform.

Related Biases: Stereotype Threat, where negative stereotypes about a person's social group can hinder their performance, and Confirmation Bias, the tendency to search for, interpret, and recall information that confirms pre-existing beliefs.

Countermeasures: Encouraging individuals to recognize their unique strengths and not to rely solely on group stereotypes can help. Educating people about the dangers of stereotypes—even positive ones—is essential.

Key Takeaways: Stereotype Boost and Lift remind us that stereotypes can influence behavior, but this isn't always a good thing. As individuals, we should strive to see past societal labels and foster personal growth and understanding.

References:

Shih, M., Pittinsky, T. L., & Ambady, N. (1999). Stereotype Susceptibility: Identity Salience and Shifts in Quantitative Performance. Psychological Science, 10(1), 80-83.

Walton, G. M., & Cohen, G. L. (2003). Stereotype Lift. Journal of Experimental Social Psychology, 39(5), 456-467.

Stereotype Threat

Definition: The Stereotype Threat refers to the psychological predicament where individuals fear that their actions may confirm existing stereotypes about their social group, thereby potentially influencing their performance and behavior.

Description: This unique twist on social anxiety puts the human brain into a fun little knot, like trying to solve a Rubik's Cube in the dark, wearing mittens, on a roller coaster. The Stereotype Threat is not about the stereotypes themselves, but the fear of confirming them. This fear, once triggered, can impact a person's cognitive resources and motor functions, much like trying to drive a car while nervously keeping an eye on a spider crawling up your leg.

Examples: Imagine you are a fervently religious individual who got invited to a science symposium. As you mingle with scientists, you fear that you may confirm the stereotype that religious individuals reject scientific concepts, causing you to overcompensate, and end up alienating your listeners by constantly asserting your acceptance of quantum physics.

Similarly, consider a left-wing politician attending a farmers' convention. Aware of the stereotype that liberal politicians are disconnected from rural life, they might take great pains to prove otherwise. They could even attempt to ride a bull to show their "rural chops," only to affirm a different stereotype that city dwellers can't handle farm life.

Impact: The Stereotype Threat has a notable influence on decision-making, judgment, and perception. It can lead to decreased performance, increased anxiety, and even self-fulfilling prophecies. This can create an unfair disadvantage in various domains, ranging from academics to politics and social interactions, not unlike trying to play a piano with one hand tied behind your back and a yodeling walrus for a music teacher.

Related Biases: Stereotype Threat is intimately linked to biases such as Self-Fulfilling Prophecy (behaving in ways that confirm our own or others' expectations), and Confirmation Bias (favoring information that confirms our pre-existing beliefs).

Countermeasures: One way to combat Stereotype Threat is through promoting positive stereotypes and creating environments that minimize stereotype activation. Think of it as flipping the script: Instead of fearing the stereotype, you use it as a springboard to break those preconceived notions. Empowerment through education and awareness can also help individuals understand and manage the threat.

Key Takeaways: Stereotype Threat is a psychological minefield that can affect anyone belonging to any social group. It's not about the stereotype, but the fear of confirming it. Understanding and recognizing this bias can help alleviate its effects, much like knowing a bear is less likely to eat you if you don't run away screaming (although, running might seem like a pretty good idea).

References:

Steele, C. M., & Aronson, J. (1995). Stereotype threat and the intellectual test performance of African Americans. Journal of personality and social psychology, 69(5), 797.

Schmader, T., Johns, M., & Forbes, C. (2008). An integrated process model of stereotype threat effects on performance. Psychological review, 115(2), 336.

Stereotyping

Definition: Stereotyping is the process of categorizing and assigning traits to individuals or groups based on preconceived notions, akin to deciding that all mushrooms are poisonous just because one gave you a terrible bout of food poisoning.

Description: This cognitive bias is the mental equivalent of buying a "one-size-fits-all" t-shirt and finding out it fits absolutely no one correctly. It involves overgeneralization and assumption-making that overlooks the rich variety of individual characteristics in favor of simplified, often erroneous group labels. Just as you wouldn't judge a book by its cover, you shouldn't judge people by their demographic group. But alas, the human brain does have a knack for taking shortcuts.

Examples: For instance, let's consider a scenario where you attend a political rally. You spot a person wearing a cowboy hat and immediately categorize him as a conservative, ignoring the fact that the hat could simply be a fashion choice or a protection against the scorching sun.

Similarly, imagine a devoutly religious person attending an atheist conference. They may stereotype the attendees as immoral or nihilistic, overlooking the fact that morality and existential purpose can be derived from sources other than religion. In this case, the religious person might be surprised to find out that the atheist crowd donates more to charity than their religious group.

Impact: Stereotyping can have substantial effects on decision-making, judgment, and perception, leading to discrimination, bias, and even conflict. It's as if our brains are playing a game of bingo where the prize is misunderstanding and social disharmony.

Related Biases: Stereotyping is closely tied to other cognitive biases like Outgroup Homogeneity Bias (perceiving members of an out-group as more similar than they are), and Confirmation Bias (favoring information that confirms our pre-existing beliefs).

Countermeasures: Overcoming stereotyping involves fostering awareness and encouraging individual assessment over generalized assumptions. It's like taking off those tacky 3D glasses and seeing the world for what it is, not what our biases would have us believe.

Key Takeaways: Stereotyping is a mental shortcut that leads to overgeneralization and incorrect assumptions about individuals or groups. While it might seem as unavoidable as a sneeze, with awareness and a commitment to see individuals for who they are rather than the groups they belong to, we can certainly mitigate its impacts.

References:

Fiske, S. T., & Taylor, S. E. (1991). Social cognition (2nd ed.). New York: McGraw-Hill.

Hamilton, D. L., & Trolier, T. K. (1986). Stereotypes and stereotyping: An overview of the cognitive approach. In J. F. Dovidio & S. L. Gaertner (Eds.), Prejudice, discrimination, and racism (pp. 127-163). Orlando, FL, US: Academic Press.

Allport, G. W. (1954). The nature of prejudice. Reading, MA: Addison-Wesley.

Status Quo Bias

Definition: Status Quo Bias refers to the cognitive inclination to prefer the current state of affairs over any change, much like opting for a third bowl of so-so mac and cheese over trying that exotic, possibly mind-blowing, cuisine.

Description: This bias is essentially our mind's way of sticking with the devil it knows, as opposed to the devil it doesn't. It arises from our inherent aversion to loss, the comfort we find in familiarity, and our reluctance to put effort into change. This bias whispers in our ear that "change is dangerous," and "let's just stick to what we've got" even when what we've got resembles a dilapidated shack more than a comfortable abode.

Examples: Politically, status quo bias can be seen when voters consistently re-elect ineffective officials, adhering to the "better the devil you know" philosophy.

In the realm of religion, it manifests when individuals, aware of the contradictions in their belief system, still cling to their traditional faith, because exploring new philosophical terrains would mean venturing into the unknown, and who has the time or energy for that?

On social issues, it can cause people to resist changes to outdated societal norms or laws because, well, "we've always done it this way."

Impact: The impact of Status Quo Bias is evident in decision-making and judgment, often leading to resistance to progress, stagnation, and the potential missed opportunities of positive change. In other words, we might be missing out on the party of the century because we're too busy wallowing in our comfort zone.

Related Biases: The Status Quo Bias is closely related to Loss Aversion (preferring to avoid losses over acquiring equivalent gains), Endowment Effect (placing higher value on things we already own), and Choice-Supportive Bias (the tendency to retroactively attribute positive features to an option we've selected).

Countermeasures: Combating Status Quo Bias involves conscious efforts to consider alternatives and assess the potential gains from change. It's about recognizing that the "unknown" is not necessarily a den of monsters but might instead be a gateway to improvement and growth.

Key Takeaways: The Status Quo Bias nudges us to prefer the familiar and resist change. While this might offer a sense of comfort, it can also result in stagnation and missed opportunities. Recognizing this bias allows us to give the new and different a fair chance, much like trying a new flavor of ice cream and discovering a new favorite.

References:

Samuelson, W., & Zeckhauser, R. (1988). Status quo bias in decision making. Journal of risk and uncertainty, 1(1), 7-59.

Kahneman, D., Knetsch, J. L., & Thaler, R. H. (1991). Anomalies: The endowment effect, loss aversion, and status quo bias. The journal of economic perspectives, 5(1), 193-206.

Streetlight Effect

Definition: The Streetlight Effect is a cognitive bias that causes people to search for answers where it's easiest to look rather than where the truth is likely to be hiding. Imagine looking for your lost keys under a streetlight, not because you lost them there, but because it's easier to see there.

Description: This bias is the cognitive equivalent of someone sticking their head in the sand and assuming that because they can't see the problem, it must not exist. It arises from our penchant for simplicity and our natural aversion to effortful thinking. It's like choosing to jog on a smooth path rather than a rocky one, even though the rocky one may lead to the finish line.

Examples: In the arena of politics, the Streetlight Effect could manifest when politicians focus on easily measurable issues, like budget deficits, rather than more complex, but perhaps more important issues, like climate change or societal inequality.

In terms of religion, it can be seen when adherents focus on surface-level rituals rather than delving into the underlying ethical or moral teachings. It's easier to light a candle and chant a hymn than wrestle with philosophical dilemmas.

In social issues, it may cause us to focus on overt instances of racism or sexism, while ignoring the more subtle, systemic issues, because it's simpler to point a finger at a villainous individual than to dissect an entrenched system.

Impact: The Streetlight Effect leads to incomplete understandings, faulty decision-making, and can often lead us astray from the truth. It's like trying to complete a jigsaw puzzle with half the pieces missing. Sure, you might end up with something that vaguely resembles the picture on the box, but there's going to be a lot of gaps.

Related Biases: The Streetlight Effect is associated with the Availability Heuristic (relying on immediate examples when evaluating a specific topic, concept, method or decision), and Confirmation Bias (favoring information that confirms our pre-existing beliefs).

Countermeasures: Countering the Streetlight Effect involves being willing to venture into the "dark" areas of understanding – the complex, the nuanced, the challenging. It's like opting to explore the rest of the iceberg beneath the surface, rather than being content with what's visible above the water.

Key Takeaways: The Streetlight Effect nudges us towards easy answers at the cost of the truth. Recognizing this bias can push us to dig deeper and challenge ourselves to seek the truth where it actually lies, not just where it's easiest to look.

References:

Kaplan, R. M., & Irvin, V. L. (2015). Likelihood of Null Effects of Large NHLBI Clinical Trials Has Increased over Time. PLoS ONE, 10(8), e0132382.

Von Hippel, W., & Trivers, R. (2011). The evolution and psychology of self-deception. Behavioral and Brain Sciences, 34(1), 1–16.

Subadditivity Effect

Definition: The Subadditivity Effect is a cognitive bias that has us believe that the sum of the parts is less than the whole. Imagine going to a buffet and thinking that one slice from each of the ten different pizzas is less calorific than eating an entire pizza by itself. Dream on!

Description: The Subadditivity Effect is essentially our brain's version of bad math. It arises when we estimate probabilities. If we're asked to estimate the probability of a broad category and then of specific components of that category, we tend to assign higher probabilities to the individual components, which would add up to more than the estimated probability for the entire category.

Examples: In politics, the Subadditivity Effect may come into play when estimating election results. Voters may assign high probabilities to multiple candidates winning, even though those probabilities would exceed 100% when combined.

In the realm of religion, believers may assign high probabilities to multiple divine interventions occurring, even if those would exceed their belief in divine intervention in general.

In social issues, we might see the Subadditivity Effect when people overestimate the prevalence of multiple individual types of discrimination, even if those estimates would exceed their perceived prevalence of discrimination overall.

Impact: This bias can lead to overestimation, miscalculations, and a skewed perception of reality, not unlike the disappointment of stepping onto the scales after indulging at that buffet. It can also affect decision-making, from resource allocation to risk assessment.

Related Biases: The Subadditivity Effect shares some features with the Conjunction Fallacy (the incorrect belief that specific conditions are more probable than a single general one) and the Disjunction Fallacy (underestimating the probability of the occurrence of at least one of two events).

Countermeasures: To mitigate this bias, one needs to stay aware of the laws of probability and the risk of overestimation. A dash of critical thinking and a smidgen of skepticism can also help keep our calculations in check.

Key Takeaways: The Subadditivity Effect can lead us astray in estimating probabilities, causing overestimation and miscalculations. Recognizing this bias helps in making more accurate predictions and decisions, or at least, not fooling ourselves about the calorie count at the buffet.

References:

Tversky, A., & Koehler, D. J. (1994). Support theory: A nonextensional representation of subjective probability. Psychological review, 101(4), 547.

Fox, C. R., & Rottenstreich, Y. (2003). Partition priming in judgment under uncertainty. Psychological science, 14(3), 195-200.

Subjective Validation

Definition: Subjective Validation, also known as the Forer effect or Barnum effect, is a cognitive bias where individuals believe vague and general statements to be uniquely applicable to themselves. Think horoscopes saying, "You will face a challenge today," and you immediately thinking about that unread email lurking in your inbox.

Description: This bias occurs because we humans love to feel special, unique, and understood. When we find personal meaning in otherwise random or generic statements, we fall for subjective validation. It's like accepting a cookie-cutter compliment for a job well done when, in reality, you spent the whole day scrolling through cat videos.

Examples: In politics, subjective validation can be seen when politicians make sweeping promises like, "I'll create jobs," and voters feel as if it's a personal commitment to them.

Religion often utilizes the bias through broad moral or prophetic statements that believers find personally relevant. "The meek shall inherit the earth" suddenly makes you feel very good about not voicing your opinion in that last Zoom meeting.

In social issues, subjective validation might manifest when individuals attribute broad societal trends to their personal experiences, reinforcing their beliefs or prejudices.

Impact: Subjective validation can lead to misinterpretations, false beliefs, and, in extreme cases, exploitation. It can cause us to believe in the accuracy of pseudoscience, fortune-telling, or propaganda. It's the "everyone gets a medal" of cognitive biases, where we see personal significance in every general statement.

Related Biases: This bias is linked with Confirmation Bias (favoring information that confirms our pre-existing beliefs) and the Illusory Correlation (perceiving a relationship between variables even when none exists).

Countermeasures: To combat this bias, skepticism and critical thinking are essential. Ask yourself, "Could this statement apply to anyone else?" If the answer is yes, then you might be falling for the charm of subjective validation.

Key Takeaways: Subjective Validation can lead us to see personal relevance in general statements, reinforcing our beliefs and sometimes leading to misinterpretation or exploitation. Being aware of this bias helps us navigate information more objectively.

References:

Forer, B.R. (1949). The fallacy of personal validation: A classroom demonstration of gullibility. Journal of Abnormal and Social Psychology, 44, 118–123.

Dickson, D. H., & Kelly, I. W. (1985). The 'Barnum effect' in personality assessment: A review of the literature. Psychological reports, 57(2), 367-382.

Survivorship Bias

Definition: Survivorship bias refers to the logical error of concentrating on the people or things that "survived" some process and inadvertently overlooking those that didn't because of their lack of visibility. It's like hosting a family reunion and then assuming the entire family is in good health, conveniently ignoring the fact that those who were sick couldn't attend.

Description: This bias often leads to false conclusions in several different ways. It can lead to overoptimistic beliefs because failures are ignored. The irony is that the cemetery of failed attempts can provide as much—if not more—insight as the success stories we idolize.

Examples: Survivorship bias can be found in many areas, such as the perception of entrepreneurial success (we focus on the Gates and the Jobs of the world, not the countless unseen failures), political narratives (highlighting the triumphant underdog stories while neglecting those who didn't make it), or religious miracles (counting the hits and ignoring the misses).

Impact: By focusing only on success or survival, we can miss essential truths and lessons. This bias can distort our understanding of reality and lead to poor decision-making, risky behavior, or unrealistic expectations.

Related Biases: Availability bias, where people rely on immediate examples that come to mind when evaluating a topic or decision, can often play into Survivorship Bias.

Countermeasures: Critical thinking and investigation are vital to combat survivorship bias. Ask: "What am I not seeing?" "What voices aren't being heard?" and "What information might be missing?"

Key Takeaways: Survivorship bias blinds us to the unseen failures or hardships. We must actively seek out the silent, the missing, and the unsuccessful to understand the full picture and make balanced decisions.

References:

Kaplan, S., & Garrick, B. J. (1981). On The Quantitative Definition of Risk. Risk Analysis, 1(1), 11-27.

Malmendier, U., & Tate, G. (2005). CEO Overconfidence and Corporate Investment. The Journal of Finance, 60(6), 2661-2700.

Sunk Cost Fallacy

Definition: The sunk cost fallacy refers to the tendency to follow through on an endeavor if one has already invested time, effort, or money into it, whether or not the current costs outweigh the benefits. It's like eating that fifth slice of pizza just because you've paid for it, even though your stomach is pleading with you to stop.

Description: This bias can make us irrational, stubborn, and wasteful. It often emerges from our inherent aversion to loss and waste. We tend to think that the more we invest in something, the more we "own" it, and the harder it is to let it go. And yet, the crust doesn't get any less tough just because we're attached to it.

Examples: This bias often creeps into politics where parties stick with failing policies just because they've invested heavily in them (think "too big to fail"). It can also influence religious attitudes—when one has devoted years to a belief, it's harder to leave, even in the face of contradicting evidence.

Impact: The sunk cost fallacy can lead to poor decision-making and wasteful investments. It can perpetuate unprofitable situations, unfulfilling relationships, or harmful beliefs, by keeping us stuck in a cycle of investment and loss.

Related Biases: Loss aversion, where the pain of losing is psychologically about twice as powerful as the pleasure of gaining, often contributes to the sunk cost fallacy.

Countermeasures: Regularly re-evaluating decisions based on current and future value, rather than past investments, can help. Embrace failure as an option and understand that sunk costs are just that—sunk. They're gone, and no amount of wishing will bring them back.

Key Takeaways: The sunk cost fallacy keeps us holding onto the past. Recognize that costs spent cannot be recovered and focus instead on the potential future benefits and losses when making decisions.

References:

Arkes, H. R., & Blumer, C. (1985). The Psychology of Sunk Cost. Organizational Behavior and Human Decision Processes, 35(1), 124-140.

Staw, B. M. (1976). Knee-deep in the big muddy: A study of escalating commitment to a chosen course of action. Organizational Behavior and Human Performance, 16(1), 27-44.

System Justification

Definition: System Justification is the bias where people tend to defend and uphold the status quo, even at the expense of individual and group interests. This is the mental equivalent of that famous line from every mediocre horror movie ever, "Let's split up, gang."

Description: We like to think that our social, economic, and political systems are just and fair, because the alternative—that we're living in an unjust world—can be too discomforting. System justification can often lead us to defend the system that disadvantages us, a phenomenon which psychologists lovingly refer to as "shooting oneself in the foot syndrome" (not really, but they should).

Examples: In politics, this might be seen when individuals from disadvantaged socioeconomic groups support tax policies that favor the rich. In the realm of social issues, victims of prejudice may sometimes defend the societal norms that perpetuate that prejudice.

Impact: System Justification can maintain and exacerbate social inequality and can hinder progress towards more equitable societies. This bias can stifle critical thought, impede social reform, and perpetuate harmful systems and norms.

Related Biases: Status Quo Bias and the Just-World Hypothesis often go hand-in-hand with System Justification.

Countermeasures: Fostering critical thinking, encouraging diversity of perspectives, and advocating for fairness and social justice can help to counteract System

Justification. Remember: challenging the status quo isn't a crime—it's often a necessity.

Key Takeaways: System Justification can blind us to the flaws in our social systems. It's important to question, scrutinize, and challenge societal norms to promote more equitable and just systems.

References:

Jost, J. T., Banaji, M. R., & Nosek, B. A. (2004). A Decade of System Justification Theory: Accumulated Evidence of Conscious and Unconscious Bolstering of the Status Quo. Political Psychology, 25(6), 881-919.

Kay, A. C., & Jost, J. T. (2003). Complementary Justice: Effects of "Poor but Happy" and "Poor but Honest" Stereotype Exemplars on System Justification and Implicit Activation of the Justice Motive. Journal of Personality and Social Psychology, 85(5), 823-837.

Telescoping Bias

Definition: Telescoping Bias is the cognitive pitfall that leads us to misremember the timing of events—specifically, it's our tendency to perceive recent events as having occurred further in the past, and distant events as having occurred more recently. If you've ever found yourself saying, "Wow, 2020 feels like it was just yesterday!" in the middle of 2023, you've got Telescoping Bias to thank.

Description: Telescoping Bias is sort of like the funhouse mirror of memory; it distorts our perception of when events occurred. This happens due to the quirks in our mental timekeeping, which is about as reliable as a sundial in Seattle. Instead of being a faithful chronology, our memory is more of an imaginative artist with a dramatic flair for reshuffling past events.

Examples: In politics, this bias might manifest as voters remembering political scandals or achievements as more recent than they actually are, leading to skewed perceptions of candidates. In the realm of religion, adherents may telescope the timeline of their religious experiences, distorting the sequence and significance of events.

Impact: Telescoping Bias can distort our perception of history, risk, and personal experience, impacting everything from our voting behavior to our recollection of personal achievements and failures. It's the reason we say time flies when we're having fun and drags when we're not, despite time's stubborn insistence on ticking at a constant rate.

Related Biases: This bias is related to the Availability Heuristic and Hindsight Bias, two other delightful ways our memories love to play tricks on us.

Countermeasures: Keeping a journal or timeline can help to counteract this bias by providing a more objective record of when events actually occurred. But beware: No amount of journaling will make 2020 feel like it was any further away.

Key Takeaways: Our perception of time is a bit wonky—future events seem far off, past events seem either too close or too far, depending on when they actually happened. To keep telescoping bias at bay, try keeping a record of important events.

References:

Loftus, E. F. (2005). Planting misinformation in the human mind: A 30-year investigation of the malleability of memory. Learning & Memory, 12(4), 361-366.

Hasher, L., Goldstein, D., & Toppino, T. (1977). Frequency and the conference of referential validity. Journal of Verbal Learning and Verbal Behavior, 16(1), 107-112.

Temporal Discounting

Definition: Temporal Discounting is the tendency to devalue rewards as they move further into the future. It's the cognitive equivalent of "a bird in the hand is worth two in the bush." If you've ever chosen immediate gratification over a potentially larger reward in the future, congratulations, you're human.

Description: Our brains are wired to favor immediate rewards, even if waiting would lead to a better outcome. This is Temporal Discounting in a nutshell. Think of it as financial impatience. It's the reason why we're more likely to take $50 today rather than $100 in a year. Apparently, our mental accounting system didn't get the memo about inflation.

Examples: In politics, this could manifest as voters favoring policies with immediate benefits, rather than those that promise greater long-term gains. For religion, it could be choosing immediate pleasures over potential eternal rewards (no, not those rewards).

Impact: Temporal Discounting can lead to suboptimal decision-making, contributing to behaviors like procrastination, poor financial planning, and overconsumption of unhealthy foods. It's essentially the reason why your future self is always mad at your past self.

Related Biases: This bias is closely related to the Present Bias, Hyperbolic Discounting, and the Instant Gratification Bias, because we just love immediate rewards that much.

Countermeasures: One way to mitigate this bias is by employing the concept of "future self-continuity," where you vividly imagine your future self. It's a bit like Scrooge meeting the Ghost of Christmas Yet to Come, only less terrifying.

Key Takeaways: Temporal Discounting is the reason we often choose immediate rewards over larger, future ones. To keep this bias in check, consider your future self when making decisions. Just remember, that future yacht won't buy itself.

References:

Frederick, S., Loewenstein, G., & O'Donoghue, T. (2002). Time discounting and time preference: A critical review. Journal of Economic Literature, 40(2), 351-401.

Benhabib, J., Bisin, A., & Schotter, A. (2010). Present-bias, quasi-hyperbolic discounting, and fixed costs. Games and Economic Behavior, 69(2), 205-223.

The Texas Sharpshooter Fallacy

Definition: The Texas Sharpshooter Fallacy occurs when one focuses on the similarities in data to the exclusion of the differences, cherry-picking data clusters based on their correlation to outcomes. It's like shooting random bullet holes and then painting a bullseye around them to showcase your "accuracy".

Description: This fallacy is an error in perception and analysis where patterns are identified in data despite the lack of actual correlation or causation. Essentially, it's trying to force a square peg into a round hole and then proclaiming, "See, perfect fit!"

Examples: In politics, one might use this fallacy to show how the implementation of a specific policy coincides with positive economic outcomes, conveniently ignoring any negative aspects or unrelated positive influences. In religion, it can be seen in the interpretation of sacred texts, where disparate passages are linked to create a narrative that supports a specific viewpoint.

Impact: This fallacy can lead to faulty conclusions, promoting misinformation and fostering bias. It's like fool's gold for decision-making, making things appear valuable or significant when they are not.

Related Biases: This fallacy is closely related to Confirmation Bias, the Clustering Illusion, and the Post Hoc Fallacy. All are essentially cognitive line-dances that can lead us astray.

Countermeasures: Promote critical thinking and statistical understanding. And when you see a bullseye, make sure it was there before the bullets were fired.

Key Takeaways: The Texas Sharpshooter Fallacy is a trickster, making us see connections where none exist. Remember, correlation doesn't equal causation, and not every pattern signifies meaningful information.

References:

Tversky, A., & Kahneman, D. (1971). Belief in the law of small numbers. Psychological Bulletin, 76(2), 105.

Chapman, L. J., & Chapman, J. P. (1967). Genesis of popular but erroneous psychodiagnostic observations. Journal of Abnormal Psychology, 72(3), 193.

Third Person Effect

Definition: The Third Person Effect is a cognitive bias where people believe that others are more influenced by persuasive communications, such as media messages, than they themselves are. In short, it's the mental gymnastics we perform to convince ourselves that everyone else is a pushover but not us.

Description: Named by sociologist W. Phillips Davison in 1983, the Third Person Effect proposes that when it comes to persuasion, we see ourselves as mental fortresses while others are seen as putty. This self-enhancing bias often leads to an underestimation of a message's effect on oneself and an overestimation of its effect on others.

Examples: The Third Person Effect is found in many social issues and political campaigns. For instance, many believe that political propaganda affects the masses but not them, leading to a self-exemption from fact-checking. Likewise, religious followers often believe that others are more prone to being led astray by contrary beliefs, while they themselves are unaffected.

Impact: This bias can cause people to dismiss relevant and valuable information, leading to poorly informed decisions. It's like having earplugs in during a concert and then complaining about the sound quality.

Related Biases: The Third Person Effect is related to the Optimistic Bias, Illusion of Invulnerability, and Bias Blind Spot. Each of these biases makes us believe we're somehow immune to the weaknesses of mere mortals.

Countermeasures: Self-awareness is key. Acknowledge that you, too, can be influenced and work on critical thinking skills. It's not admitting defeat; it's embracing humility.

Key Takeaways: The Third Person Effect is a reminder that we're not as impervious as we'd like to believe. Accepting our susceptibility to persuasion is the first step towards more informed decisions.

References:

Davison, W. P. (1983). The third-person effect in communication. Public Opinion Quarterly, 47(1), 1-15.

Perloff, R. M. (1993). The dynamics of persuasion. Lawrence Erlbaum Associates, Inc.

Thought Suppression

Definition: Thought Suppression is a cognitive bias where attempts to repress or ignore certain thoughts ironically make those thoughts more prevalent. It's like telling you not to think about pink elephants; suddenly, pink elephants become the star of your mental show.

Description: This psychological phenomenon, commonly known as the 'White Bear Problem', is named after a famous experiment by Wegner et al., 1987. They instructed participants not to think about a white bear and - wouldn't you know it - the white bear was all they could think about. This can occur with thoughts about anything from a catchy tune to more substantial concerns, like political issues or religious doubts.

Examples: In the realm of politics, suppressing thoughts about a controversial policy could result in thinking more about it, affecting one's voting behavior. Similarly, in religion, trying to suppress doubts might lead to more questions and potential upheaval in faith.

Impact: Thought Suppression can lead to an increased focus on unwanted thoughts, causing distress, impaired decision-making, and a potential spiral into negative thinking. It's like trying to put out a fire with gasoline – it just doesn't end well.

Related Biases: Thought Suppression is related to the Ironic Process Theory and the Rebound Effect. These biases highlight the irony of our minds, which seem to have a sense of humor when it comes to doing exactly the opposite of what we want.

Countermeasures: Rather than attempting to suppress thoughts, acknowledge them, and let them pass naturally. Mindfulness and cognitive behavioral techniques can be particularly helpful. Remember, the brain is a wild beast, and attempting to tame it often leads to more chaos.

Key Takeaways: Trying to suppress thoughts can have the opposite effect, increasing their prevalence. Accepting and acknowledging our thoughts, rather than trying to push them away, can lead to better mental control.

References:

Wegner, D. M., Schneider, D. J., Carter, S. R., & White, T. L. (1987). Paradoxical effects of thought suppression. Journal of personality and social psychology, 53(1), 5.

Wenzlaff, R. M., & Wegner, D. M. (2000). Thought suppression. Annual review of psychology, 51(1), 59-91.

Time-saving Bias

Definition: Time-saving Bias is the cognitive bias that convinces us that time-saving methods will save more time than they actually do. It's like buying a new gadget that promises to save you an hour a day but conveniently forgetting that you'll spend half of your life figuring out how to use it.

Description: This bias stems from our optimistic prediction of how much time we can save by changing our current processes or habits. We overestimate the benefits of time-saving methods, often overlooking the fact that they have their own set of time-consuming elements. And you thought you were gaming the system, didn't you?

Examples: In politics, a new policy or software may be promoted as time-saving, yet the hours spent in training or the effort of implementation may outweigh the benefits. Similarly, in religion, adopting new forms of worship to save time may not account for the time spent learning and adapting to the new methods.

Impact: Time-saving Bias can lead to poor decisions about resource allocation, such as money or effort, based on overoptimistic assumptions. It's as if we're always expecting time to bend to our will, only to find out it's not quite as flexible as we'd like.

Related Biases: This bias is related to the Planning Fallacy, where we underestimate the time it takes to complete tasks. They both show our fantastic, yet misguided, ability to be eternal optimists when it comes to time management.

Countermeasures: Consider the full context and all the time-related factors when evaluating time-saving methods. This includes the time needed to learn, implement, and possibly troubleshoot the new system. A bit of realism can go a long way in saving you from the alluring trap of the time-saving bias.

Key Takeaways: While time-saving methods can be beneficial, it's crucial to consider all factors, including the time to implement and learn the new method. Don't let the prospect of a free hour tomorrow fool you into wasting three hours today.

References:

Chung, T. M., & Han, I. (2014). Time-saving bias in users' e-learning system continuance intention under high perceived pressure: A moderating role of time management disposition. Behaviour & Information Technology, 33(7), 733-747.

Moon, Y. J., & Kim, Y. G. (2001). Extending the TAM for a World-Wide-Web context. Information & Management, 38(4), 217-230.

Trait Negativity Bias

Definition: Trait Negativity Bias is the cognitive glitch that compels us to weigh negative traits more heavily than positive ones. When it comes to our perception of others, we're like unenthusiastic movie critics, focusing more on the one bad scene rather than the many good ones.

Description: This bias is often seen when we judge others based on their actions or characteristics. Our brains have a tendency to spotlight the negatives, pushing the positives backstage. Yes, we are all closet pessimists.

Examples: In politics, a candidate's small misstep might overshadow their numerous accomplishments. Or in religion, a single point of doctrinal disagreement may cause one to overlook shared beliefs. In social issues, people might dwell more on a person's negative characteristics, like being late, rather than their positive traits, like being hard-working.

Impact: The Trait Negativity Bias can skew perception, hinder effective decision-making, and lead to unfair judgments. On a larger scale, it can fuel conflict in political, religious, or social contexts. No wonder why the world is such a mess.

Related Biases: This bias is related to the Negativity Bias, where we tend to focus more on negative events than positive ones, and the Halo Effect, where one positive trait can influence our overall impression of a person.

Countermeasures: Promoting awareness of this bias can help mitigate its effects. Attempt to consider the whole person or situation, not just the negative traits that stand out. You know, just like mom told you to do.

Key Takeaways: People are complex bundles of traits, both good and bad. Overemphasizing negative traits can lead to unfair judgments and unbalanced perceptions. Don't let the darkness block out the light.

References:

Peeters, G., & Czapinski, J. (1990). Positive-negative asymmetry in evaluations: The distinction between affective and informational negativity effects. European Review of Social Psychology, 1(1), 33-60.

Rozin, P., & Royzman, E. B. (2001). Negativity bias, negativity dominance, and contagion. Personality and Social Psychology Review, 5(4), 296-320.

Trivial-Pursuit Effect

Definition: The Trivial-Pursuit Effect refers to the tendency for people to accumulate unnecessary and insignificant facts, information, or skills that may not be useful in meaningful or practical contexts. Essentially, it's like cramming your brain with every factoid you could find just in case you ever land on a game show.

Description: This cognitive bias results from our fascination with quirky, unusual, or otherwise trivial knowledge. The novelty of such information can make it seem more valuable or important than it really is. It's like being the guy at the party who can name all the state capitals – impressive, but unlikely to be useful unless you're at a very specific kind of party.

Examples: For example, in politics, one might focus on insignificant details about a candidate's personal life rather than their policy proposals. In religion, people might become preoccupied with obscure theological trivia while missing the broader teachings and principles. In social issues, attention may be drawn to sensational but inconsequential aspects, overshadowing the fundamental problems at hand.

Impact: The Trivial-Pursuit Effect can divert our attention from what is truly important and practical. It might make you the king or queen of trivia night, but in more serious contexts, it can lead to poor decision-making and unproductive conversations.

Related Biases: This effect is akin to the Red Herring Fallacy, where irrelevant information is used to distract from the real issue at hand.

Countermeasures: Stay focused on what's important. Challenge yourself to sift through the noise and prioritize information that's genuinely significant and relevant. In essence, don't study for a life that's a game show.

Key Takeaways: We often give undue attention to trivial information, which can lead to misdirected focus and less-than-ideal decisions. Stay tuned into what really matters.

References:

Alba, J. W., & Hutchinson, J. W. (2000). Knowledge calibration: What consumers know and what they think they know. Journal of Consumer Research, 27(2), 123-156.

Bialik, M., & Fadel, C. (2018). What we say and what we do: reflexivity, emotions and power in children and young people's participation. International Journal of Qualitative Studies in Education, 31(3), 202-217.

Triviality Bias

Definition: The Triviality Bias refers to the cognitive bias where people give undue attention to trivial issues while ignoring significant ones. Think of it as the mental equivalent of rearranging deck chairs on the Titanic.

Description: This bias can be attributed to various psychological factors, including decision fatigue, overwhelm, or even a desire to avoid complex problems. Trivial issues are simpler, easier, and less stressful to tackle. As such, the Triviality Bias is often a procrastinator's best friend and an efficiency enthusiast's worst enemy.

Examples: Politically, this bias might manifest when citizens focus more on a candidate's hair style rather than their stance on key policy issues. In religious discussions, minor doctrinal differences may become the focal point rather than the core beliefs that unite the community. On social matters, debates may devolve into arguments about terminology rather than tackling the underlying social issues at hand.

Impact: The Triviality Bias can lead to inefficiency, wasted time, and a failure to address pressing concerns. It's like spending all your time arguing about what color to paint the parachute, forgetting the fact that you're free-falling without one.

Related Biases: The Triviality Bias is closely related to Parkinson's Law of Triviality, also known as bikeshedding, and Majoring in Minors, where disproportionate attention is given to trivial issues.

Countermeasures: To counter this bias, focus on priorities, use structured decision-making processes, and encourage discussions that tackle substantive matters. In other words, don't get lost in the sauce of minor details when there's a whole gourmet meal of significant issues to address.

Key Takeaways: The Triviality Bias can cause us to focus on trivial issues at the expense of significant ones. To make effective decisions and discussions, give due attention to what truly matters.

References:

Parkinson, C. N. (1957). Parkinson's law, and other studies in administration. Houghton Mifflin.

Conlon, D. E., & Garland, H. (1993). The role of project completion information in resource allocation decisions. Academy of Management Journal, 36(2), 402-413.

Ultimate Attribution Error

Definition: The Ultimate Attribution Error is the cognitive bias where individuals attribute negative behaviors of out-group members to their inherent nature (dispositional) and positive behaviors to environmental circumstances (situational), while the opposite is true for in-group members.

Description: This bias is the hot sauce on the burrito of our biases, it's the seasoning that gives our prejudices that extra kick. Essentially, it's an overextension of the Fundamental Attribution Error, but with a twist of group dynamics. It serves as a mental shortcut to ease cognitive dissonance when out-group members act outside of our stereotypical expectations.

Examples: Let's dive into the juicy world of politics. Picture a politician from a party you dislike doing something commendable – Ultimate Attribution Error makes it easy to dismiss their action as a fluke or politically motivated. In contrast, when a favored politician slips up, it's just a "bad day" or "media manipulation".

In the realm of religion, a member of an "other" religion doing good might be seen as an "exception to the rule," whereas negative behaviors confirm inherent "flaws" in that religion. Similarly, in social issues, it can lead to attributing the failures of marginalized groups to inherent flaws rather than systemic challenges.

Impact: The Ultimate Attribution Error can foster stereotypes, promote prejudice, and exacerbate social, religious, and political divides. It's like pouring gasoline on the fire of discrimination – pretty detrimental, if you ask me.

Related Biases: This bias is a specialized form of the Fundamental Attribution Error and is closely linked to the In-group Favoritism and Out-group Homogeneity Biases.

Countermeasures: To mitigate this bias, awareness is the first step. Recognizing the humanity and individuality of out-group members can also help. To put it bluntly, don't let your brain play the "us vs. them" game; it's a tournament you won't win.

Key Takeaways: The Ultimate Attribution Error can lead us to foster harmful stereotypes and feed social divisions. Recognize it, challenge it, and remember – people are more than the groups they belong to.

References:

Pettigrew, T. F. (1979). The ultimate attribution error: Extending Allport's cognitive analysis of prejudice. Personality and social psychology bulletin, 5(4), 461-476.

Hewstone, M. (1990). The 'ultimate attribution error'? A review of the literature on intergroup causal attribution. European Journal of Social Psychology, 20(4), 311-335.

Uncertainty Bias

Definition: Uncertainty Bias is a cognitive bias where individuals tend to avoid uncertainty, even if it may result in a potentially better outcome. This bias shows our penchant for the devil we know over the one we don't.

Description: Think of this bias as a "fear of the unknown" on steroids. We crave certainty and predictability - whether it's preferring the predictable taste of a Big Mac over the mystery of an unknown cuisine, or sticking to a miserable job rather than venturing into new career paths. Uncertainty feels like jumping off a cliff with a possibly malfunctioning parachute. And who would willingly do that?

Examples: Uncertainty Bias shows up in religion when people follow established doctrines rather than exploring alternative belief systems. In politics, it's evident when voters stick with incumbent politicians, despite not being fully satisfied, because they fear the uncertainty of change. And in social issues, people may stick to traditional norms, even if they may be discriminatory or harmful, just to avoid the uncertainty of change.

Impact: Uncertainty Bias can stifle innovation, progress, and personal growth. It can keep societies stuck in harmful traditions and prevent political reform. It's like being stuck in a bad movie because you're afraid the next one might be worse.

Related Biases: The Status Quo Bias and Loss Aversion are close siblings of the Uncertainty Bias.

Countermeasures: Recognize the bias and try to weigh potential outcomes more objectively. Embrace the fact that life is inherently uncertain and not all uncertainty is negative. In other words, train yourself to see the potentially malfunctioning parachute as an exciting opportunity for a free-fall experience.

Key Takeaways: Don't let Uncertainty Bias hold you back. Embrace the unknown, step out of your comfort zone, and remember - fortune favors the bold, not the certain.

References:

Camerer, C., & Weber, M. (1992). Recent developments in modeling preferences: Uncertainty and ambiguity. Journal of risk and uncertainty, 5(4), 325-370.

Van Dijk, E., & Zeelenberg, M. (2007). When curiosity killed regret: Avoiding or seeking the unknown in decision-making under uncertainty. Journal of Experimental Social Psychology, 43(4), 656-662.

Unit Bias

Definition: Unit Bias refers to the tendency to believe that a single unit of something - a slice of cake, a beer, a political party - is the right and proper amount, regardless of the size of that unit.

Description: Just like Goldilocks, we're obsessed with "just right" - not too little, not too much. But instead of taste-testing porridge, we apply this bias to everything from food portions to moral judgments. We view one of something as "the norm", and deviation from this arbitrary measure can feel weird, wrong, or wasteful.

Examples: In the realm of religion, the Unit Bias may lead to a belief that there's only one true religion, disregarding the diversity of belief systems worldwide. In politics, it can foster a two-party system mentality, where other political factions are seen as irrelevant or divisive. In social issues, this bias can perpetuate stereotypical norms, such as the belief in the "nuclear family" unit as the ideal.

Impact: The Unit Bias can limit our perspective and flexibility, leading us to make decisions based on unwarranted assumptions about what's 'normal' or 'appropriate'. It's like insisting on having one scoop of ice cream, even if the scoop is the size of a bowling ball.

Related Biases: The Anchoring Bias and Normalcy Bias are related to the Unit Bias in the sense that they also involve sticking to a perceived 'norm'.

Countermeasures: Becoming aware of this bias is the first step. The next is challenging our assumptions about what constitutes a 'unit' and whether it truly serves our needs or goals.

Key Takeaways: Don't let the Unit Bias limit your options. There's no one-size-fits-all solution for life, politics, religion, or ice cream. Be flexible, challenge norms, and remember – variety is the spice of life.

References:

Geier, A., Rozin, P., & Doros, G. (2006). Unit bias: A new heuristic that helps explain the effect of portion size on food intake. Psychological Science, 17(6), 521-525.

Marchiori, D., Papies, E. K., & Klein, O. (2014). The portion size effect on food intake. An anchoring and adjustment process? Appetite, 81, 108-115.

Von Restorff Effect

Definition: The Von Restorff Effect is the cognitive bias that ensures the guy at the party wearing a duck suit gets remembered, while everyone else fades into obscurity. More formally, it's the tendency to remember things that stand out or are significantly different from the rest.

Description: Named after Hedwig von Restorff, a German psychiatrist who studied this phenomenon in the 1930s, the effect highlights our propensity to remember the unusual, the distinctive, the duck-suit wearers of the world. This occurs because distinctive items capture our attention, leading to enhanced memory encoding and retrieval.

Examples: In politics, it's the politicians with unique characteristics or controversial views that tend to be most remembered. In religion, doctrines or rituals that significantly deviate from the norm are often the ones that stick in our memories. In social issues, it's the unusual stories or unique cases that make headlines and capture public attention.

Impact: While it ensures we'll always remember the quirkiest moments, the Von Restorff Effect can skew our perception of what's typical or important. It can make us pay undue attention to the loudest voices in the room, even if they don't represent the majority or the facts.

Related Biases: The Availability Heuristic and the Novelty Effect are related biases that can also skew our attention and memory towards the most distinctive or recent information.

Countermeasures: Recognize that just because something stands out, it doesn't make it more important or representative of the whole. Look for the quiet, consistent data instead of the noisy, attention-grabbing outliers.

Key Takeaways: The Von Restorff Effect reminds us that memory is not a faithful record, but a spotlight that shines brightest on the unusual. Be aware of this bias in your decision-making, and don't let the man in the duck suit unfairly dominate your recollection of the party.

References:

Von Restorff, H. (1933). Über die Wirkung von Bereichsbildungen im Spurenfeld. Psychologische Forschung, 18(1), 299-342.

Hunt, R. R. (1995). The subtlety of distinctiveness: What von Restorff really did. Psychonomic Bulletin & Review, 2(1), 105-112.

Wishful Thinking

Definition: Wishful Thinking, the cognitive bias that turns us into fairytale characters convinced that if we hope hard enough, the universe will give us what we want. Formally, it's the tendency to believe that what we desire to happen will happen, regardless of evidence to the contrary.

Description: Much like a child blowing out birthday candles with a wish, adults also fall prey to the illusion that their personal hopes and desires influence the external world. Psychologists suggest that this bias is a self-defense mechanism to maintain self-esteem and promote optimism, even if it's as grounded in reality as a unicorn on a rainbow.

Examples: In politics, wishful thinking may cause a supporter to believe in an unlikely victory, despite all polls indicating otherwise. In religion, it's seen in the conviction that performing certain rituals or making certain sacrifices will yield desired outcomes. In social issues, it might manifest as the belief that societal problems will resolve themselves without active intervention.

Impact: It's not hard to see how this cognitive equivalent of 'head-in-the-clouds' syndrome could hinder rational decision making. It can lead to overconfidence, underestimation of risks, and failure to make contingency plans. Basically, it's the cognitive bias version of "winging it".

Related Biases: The Optimism Bias and the Ostrich Effect, wherein we prefer to stick our heads in the sand rather than face uncomfortable truths, are cousins of Wishful Thinking.

Countermeasures: Start with accepting that wishing doesn't make it so. Cultivate critical thinking skills, continually challenge your assumptions, and seek evidence that contradicts your beliefs. Oh, and maybe invest in a reality check or two.

Key Takeaways: The world doesn't run on wishes and fairy dust, and while Wishful Thinking might feel good, it often leads us astray. So, ground your decisions in facts, not just in what you'd like those facts to be.

References:

Krizan, Z., & Windschitl, P. D. (2007). The influence of outcome desirability on optimism. Psychological Bulletin, 133(1), 95-121.

Babad, E. Y. (2009). Wishful thinking among voters: Motivational and cognitive influences. International Journal of Public Opinion Research, 21(1), 41-63.

WYSIATI (What You See Is All There Is)

Definition: WYSIATI, an acronym that could be the title of an unreleased ABBA song, stands for 'What You See Is All There Is'. It refers to the tendency to form conclusions based on the information available at hand, without considering the possibility of any missing data.

Description: It's essentially a case of cognitive tunnel vision. WYSIATI can lead people to ignore or underestimate relevant information they don't have, leading to an overconfidence in their conclusions. After all, why would we ever consider that there might be things we don't know we don't know?

Examples: In politics, a citizen might form an opinion about a policy based on a single news article, without seeking out further information or considering different perspectives. Similarly, in religion, one might have beliefs that are predominantly shaped by their immediate community, without considering the wider spectrum of religious thought. Social issues are also susceptible to this bias, where conclusions are often formed based on limited or skewed media coverage.

Impact: The downside of WYSIATI is that it can lead to hasty judgments, oversimplification of complex issues, and a failure to account for unknowns. The world might be simpler if what we saw was all there is, but unfortunately, it's just not the case.

Related Biases: WYSIATI is related to the Availability Heuristic, wherein we base judgments on the information that is readily available in our minds.

Countermeasures: The key to combating this bias is to constantly remind oneself that there's always more to the story. Actively seeking out additional information, particularly from diverse and contradictory sources, can help in developing a more balanced perspective.

Key Takeaways: Remember, just because you can't see it, doesn't mean it doesn't exist. Always strive to see beyond the information that's immediately available.

References:

Kahneman, D. (2011). Thinking, fast and slow. Macmillan.

Tversky, A., & Kahneman, D. (1973). Availability: A heuristic for judging frequency and probability. Cognitive Psychology, 5(2), 207-232.

Zeigarnik Effect

Definition: No, Zeigarnik isn't a word your cat walked across your keyboard and typed. The Zeigarnik Effect is a cognitive bias referring to the phenomenon that people remember uncompleted or interrupted tasks better than completed tasks.

Description: Named after the Soviet psychologist Bluma Zeigarnik who, supposedly over a cup of coffee, wondered why waiters could remember complex orders yet forget them once served. The psychology behind the effect is based on the tension that comes from unfinished work, making it more memorable. Our brains are just that into closure, it seems.

Examples: In politics, a politician promising to follow up on a contentious issue but leaving it unresolved could experience more public attention (and scrutiny) until a resolution is reached. Religiously, unresolved theological questions or paradoxes often receive more focus and engagement than settled doctrines. For social issues, a publicized case remaining unresolved tends to keep the public's attention and media spotlight, contributing to a sense of urgency.

Impact: The Zeigarnik Effect can create a mental sticking point, leading people to focus more on unresolved issues than ones they have already dealt with. While this can boost productivity and problem-solving in some cases, it can also contribute to stress, anxiety, and an overemphasis on problems rather than solutions.

Related Biases: The Zeigarnik Effect is closely related to the Ovsiankina Effect, which states that an interrupted task causes some level of discomfort, leading to an urge to complete the task and achieve closure.

Countermeasures: To make use of this effect productively, you might want to start a task to keep it on your mind, even if you can't complete it right away. To mitigate its potential downsides, practicing stress management and learning to shift focus away from unresolved problems can be beneficial.

Key Takeaways: Don't underestimate the power of unfinished business. The Zeigarnik Effect shows us that our brains are eager for resolution, and will nudge us (sometimes relentlessly) towards it. Keep this in mind when setting tasks and prioritizing your to-do list.

References:

Zeigarnik, B. (1927). Das Behalten erledigter und unerledigter Handlungen. Psychologische Forschung, 9, 1-85.

Ovsiankina, M. (1928). Die Wiederaufnahme unterbrochener Handlungen. Psychologische Forschung, 11, 302-379.

Zero Price Effect

Definition: The Zero Price Effect is a cognitive bias that suggests people often perceive the benefits associated with free products as higher than they truly are. It's the reason we collectively lose our minds when we see the word "FREE!" in shiny, bright letters.

Description: This bias exploits our irrational excitement about getting something for nothing, creating a preference for free items even when the actual value is low. It's the psychological equivalent of believing there's such a thing as a free lunch, despite countless economists rolling in their graves telling us otherwise.

Examples: In politics, offering free services or amenities can sway public opinion, even when the long-term cost is hidden. Religiously, offering free classes or events may attract more interest than those with even a small associated cost. Socially, free samples in a supermarket can draw crowds, even if the product isn't typically in high demand.

Impact: The Zero Price Effect can lead to irrational decisions, such as opting for a free item over a superior, but non-free alternative. It can cause us to consume resources unnecessarily, drive demand artificially, and distort our perception of value.

Related Biases: The Zero Price Effect is closely related to the Overvaluation Effect, where people tend to overestimate the value of their personal belongings, and the Endowment Effect, where people value items more highly simply because they own them.

Countermeasures: Cultivate the habit of evaluating "free" offers critically, considering their real cost (there's always one, trust me), and comparing their value to non-free alternatives. If an offer seems too good to be true, it's probably exploiting your Zero Price Effect bias.

Key Takeaways: Remember, just because something is free, doesn't mean it's worth the cost. The Zero Price Effect reminds us that we're all a little bit vulnerable to the lure of the "FREE!" sign. Keep your wits about you, and don't be swayed by the siren song of complimentary cheese cubes on toothpicks.

References:

Shampanier, K., Mazar, N., & Ariely, D. (2007). Zero as a special price: The true value of free products. Marketing Science, 26(6), 742-757.

Zero-Sum Bias

Definition: Zero-Sum Bias is the cognitive glitch leading us to believe that a situation is zero-sum, or win-lose, when it's actually not. In other words, it's when we think life's a giant pie, and if someone gets a bigger slice, ours must be smaller. Spoiler alert: it's not always about pie.

Description: This bias manifests when we view resources, success, and fortune as finite. It's like seeing a friend's vacation photos on social media and thinking, "That should have been me in Hawaii!" - forgetting that vacations are not a one-off commodity, and you too can experience a tropical getaway (given time and savings).

Examples: In politics, it shows up when people see immigration as a threat to their jobs. In religion, believers might think that an increase in different faith groups threatens their own faith. Socially, when your friend gets a promotion, you might see it as reducing your own chances of advancement.

Impact: Zero-Sum Bias can lead to unnecessary competition, conflict, and resentment. It can foster division in society and hinder cooperative solutions. It also makes office potlucks unnecessarily competitive (seriously, no one needs that much potato salad, Bob).

Related Biases: It's somewhat related to Loss Aversion, where the fear of losses influences decisions more than potential gains. It also aligns with the Scarcity Bias, where we place higher value on things perceived as scarce.

Countermeasures: Reflect on whether resources are truly limited or if you're seeing pies where there are none. Foster a growth mindset, recognizing that success and resources can often be expanded with effort and collaboration.

Key Takeaways: Not everything in life is a competition. Often, the only pies in question are metaphorical ones, and they can be infinitely expanded. Combat the Zero-Sum Bias by remembering that someone else's gain is not necessarily your loss.

References:

Meegan, D. V. (2010). Zero-Sum Bias: Perceived Competition Despite Unlimited Resources. Frontiers in Psychology, 1, 191.

Zero Risk Bias

Definition: Zero Risk Bias is the human preference for reducing a small risk to zero over a greater reduction in a larger risk. It's like obsessively sanitizing your hands while jaywalking across a busy street. Priorities, people.

Description: This bias reflects our irrational inclination to completely eliminate risks, even when doing so may be less effective or beneficial than reducing greater, more substantial risks. It's kind of like mopping the deck of the Titanic, post-iceberg.

Examples: Politically, it can be seen when enormous resources are diverted to remove a tiny risk completely while larger risks are left unaddressed. In social issues, people might be focused on removing the last vestiges of a minor problem, ignoring more significant systemic issues that exist.

Impact: Zero Risk Bias can lead to inefficient allocation of resources, heightened anxiety over minor issues, and neglect of more significant risks. Plus, it's the reason why we have warning labels on hair dryers about not using them in the shower.

Related Biases: This is similar to the Neglect of Probability, where people disregard probability when making a decision under uncertainty. It also relates to the Availability Heuristic, as more immediate and memorable risks are often perceived as greater.

Countermeasures: Practice risk assessment and prioritization. Ask yourself: "Is eliminating this small risk worth it, considering the larger risks I'm ignoring?" You can also channel your inner statistician and understand probabilities better.

Key Takeaways: Life involves risk, and eliminating all risks is impractical and unproductive. Focus on managing and reducing significant risks instead of chasing the illusion of a risk-free life.

References:

Pachur, T., Hertwig, R., & Steinmann, F. (2012). How Do People Judge Risks: Availability Heuristic, Affect Heuristic, or Both? Journal of Experimental Psychology: Applied, 18(3), 314–330.

Conclusion

Alright, folks, we've traveled together through the wondrous and occasionally flabbergasting maze that is the human mind, and you might have noticed a fair share of overlap and repetition in the biases we've covered. Deja vu? Nope, that's just cognitive biases for you—echoing and reechoing in their irksome, charming little ways.

This book is more than just a gripping journey through the alleyways of cognition. It's a reference manual—like a cognitive Swiss Army knife, if you will. Uncertain if you're looking at availability bias or confirmation bias? Whip out this trusty tome and cut through your confusion.

Now, you might be sitting there, wondering if we'll find a another bias tomorrow, or if some eager-beaver researchers will cast doubt on bias number 103. And you know what? They just might. The field of cognitive biases is ever-evolving, much like that plot of land your neighbor keeps gardening at odd hours. Studies come and go, results get challenged, and theories get replaced. That's just science doing its self-correcting thing, like a washing machine in a perpetual spin cycle—it's loud, it's unsettling, but it gets the job done.

As for how you wield this newfound power, well, that's up to you. Relationships, marketing, sales, persuasion, politics—the applications are as numerous as the stars in the sky, or the number of streaming services vying for your attention.

But let's not forget the main attraction of this knowledge—it's your shield and armor in this era of misinformation and disinformation. Like a literary Wonder Woman's bracelet, this book can deflect the arrows of untruths or half-truths shot your way. Remember, knowledge isn't just power—it's the force field that helps you dodge those pesky, cognitive landmines that could blast your good judgment to smithereens.

So there you have it—the big finish, the grand finale. But really, it's just the beginning. So go forth, armed with your cognitive shield, into the brave, biased world. And may the force (of rationality) be with you.

About the Author

Dr. Bo Bennett wears many hats, but his favorite is perhaps the thinking cap. With a PhD in social psychology, Dr. Bennett is a true intellectual, even though he might chuckle at such a designation. His vast contributions to the realm of critical thinking, however, speak volumes about his profound knowledge and engaging writing style.

Best known for his best-selling book on logical fallacies, "Logically Fallacious," Dr. Bennett is no stranger to tackling the complex intricacies of human cognition and reasoning. For six years, he chipped away at his latest masterpiece, "The Biased Brain." He didn't labor over it daily, mind you, but rather like a wise gardener tending his plants, he allowed it to grow and mature at its own pace.

But don't think for a moment that Dr. Bennett is a one-hit wonder. He has authored over a dozen other books, most of which dive headfirst into the riveting world of critical thinking. You can check out his literary parade on his website, BoBennett.com.

When not immersed in the labyrinth of cognitive biases and logical fallacies, Dr. Bennett manages Archieboy Holdings, LLC. As the owner of this ambitious company, he proves that his expertise stretches beyond the realm of psychology, bridging the gap between academia and the entrepreneurial world. Find out more about his ventures at archieboy.com.

Whether through his books or businesses, Dr. Bo Bennett is on a perpetual journey to instigate thought, challenge norms, and inspire rational decision-making—one cognitive bias at a time.

Notes

Made in the USA
Monee, IL
02 September 2024

65056976R00262